Soil

By

Suresh Chandra Tiwari

First published by AuthorHouse 9/14/04

ISBN: 1-4184-7829-6 (e-book)
ISBN: 1-4184-5318-8 (Dust Jacket)

Library of Congress Control Number: 2003097965

This book is printed on acid free paper.

Printed in the United States of America
Bloomington, IN

DEDICATION

To My Grandfather with Love and Gratitude

This book on 'Soil' is dedicated to the memory of my grandfather, Yashodanand Tiwari (1881-1967), a hard working individual of rural India lying on the Gangetic belt, who always spent his early day time from morning until noon working on his small farm land, with a small trowel in his hand. His only pride was to raise good crops from soil to support his joint family.

Suresh C. Tiwari

TABLE OF CONTENTS

FOREWARD

Soils have a special interest to people in many occupations – farmers, ranchers and foresters; engineers, planners and builders; scientists in many fields; sanitarians and waste disposal specialists, and almost all people have been confronted by an unruly home garden or flower bed or a difficult to manage lawn.

Many books have been written on introductory soil science for various disciplines and for specific audiences. Dr Tiwari has written a book to be read and studied by students of all ages who have little or no training on the subject but have a primary interest to learn more about this dynamic system – the soil. He has selected the subject matter to present the simpler principles with corresponding examples of practices in soil science.

Dr. Tiwari has called on his scientific and practical knowledge to design a book on introductory soil science to help students at the high school or beginning college level to develop intellectual curiosity leading toward an understanding of the basic principles underlying soil science and land use. His work is not intended to be a reference work for research soil scientists, but to provide the student with an understanding of principles and a practical approach to the solution of typical soil problems.

The text is scientific and organizational reinforcing, while the language is clear and not wordy.

H.F. "Jack" Perkins, PhD
D.W.Brooks Distinguished Professor of Agronomy Emeritus
University of Georgia

PREFACE

Most materials of this book have been gathered, over the years, from my notes, lesson-plans and lectures I prepared for and presented to the undergraduate classes I taught. It has not only been a teaching experience, but a learning experience as well. I learned in the process and recycled my learning into teaching, thereby changing and improving both. All these years, I had one question in my mind. That question was about how to make others understand and follow the information and knowledge I have had gathered and successfully used in my classrooms. In all my professional life I had worked on Soil. The spirit of the study of this subject had entered into my physical and intellectual state of being. In this Soil - book, I have made an humble attempt to make that spirit understandable to those who would be interested in learning about Soil and the materials relative to the subject. The discipline of Soil has been treated, described, and explained here as a science, so that undergraduate and/or graduate students on the subject can use it as a practical guidebook.

It is relevant at this point to tell the readers about the different phases of my personal training, learning, and understanding of Soil Science as an academic and professional discipline. These phases were: (1) My attraction to the hard working farmers in the field in Bihar, India, before entering the Agriculture College in the year 1951; (2) My undergraduate training in agriculture at Sabour (India) (1951-54); (3) My soil survey and specialist's work in Bihar (India) (1954-64); (4) My graduate studies in the University of Georgia at Athens (1964-69); and the opportunities I got to teach 'Soil' and do research in Agronomy at Alcorn State University, Lorman, Mississippi (1969-97).

I had always appreciated the inspiration I got from Dr. H.N. Mukherjee, who was my teacher in Soil Science in my undergraduate classes in India. He was instrumental in increasing my interest in the subject. Dr. Mukherjee, in his later years, worked as a soil-fertility specialist for Southeast Asia under the United Nations Food and Agricultural Organization. I have had great admiration for Drs. S.C. Mandal and P.P. Jha for exposing me to the vast discipline of Soil

Science. In this connection, I would also like to recognize Mr. S.D. Sinha, Soil Survey Officer, Bihar (India). Their continuous inspiration and guidance motivated me to get further involved in Soil Science and Agronomy. Finally, my special thanks are due for my major advisor, Dr. H.F. Perkins, Professor of Soil Science at the University of Georgia, Athens (USA), for the training he has given me. This great scholar and honorable gentleman always treated me with respect and gave me guidance and help. I do extend my heart felt thanks to Dr. T. M. Parchure of US Army Engineer Waterways Experiment Station and Mr. Anand Dwivedi; of Computer Science Department of Alcorn State University for computing, scientifically editing, correcting and arranging the entire manuscript in an orderly fashion. They also provided computer help in generating all the drawings, figures and tables included in this book.

I wish to conclude my preface with this simple and relevant truth that Soil is a non-renewable living resource, and that this precious resource should be used in a way so that it's long-term sustainability is guaranteed for the nourishment and growth of humanity. I shall consider my humble endeavor successful if the readers of this book understand this truth.

Suresh C. Tiwari

CHAPTER 1

Introduction to Soil

Soil lies in between the rocky or nonrocky crust of the lithosphere below and the atmosphere above. Besides, it is surrounded by other soils or pool of water all around. As such the location of soil in general truly represents an interface of two main diversities. This situation represents an attractive ground to invite a huge concentration of biological species which decline sharply above and below (in the media of lithosphere and atmosphere) but show gradual or abrupt changes in their surroundings where they merge into other soils or the pools of water called as oceans.

Under our foot, hiding either under grasses and forest canopy or being quite exposed to the atmosphere, this soil humbly and continuously serves enormous span of lives (microbes, plants and animals including homo-sapiens). These soil-microbes and plants not only help us grow our food, but also in many cases serve us by producing medicines and antibiotics to safeguard us from various diseases.

In addition, soil also serves as a base for huge and small structures (like buildings and roads) as well as a hiding ground for commercial waste products.

It also works as a natural filter to purify usable water and is a huge medium for consuming solar energy through its green plants for the production of biomass. These plants exist on the land as well as in ocean.

THE LAND AND THE OCEAN

When we look at the landmass, we can never ignore its immediate counterpart "ocean." These both together are the complete global arrangement of the planet "Earth." In general one third of the

Earth's surface is land and the other two third is ocean. A negligent observer may find as haphazard arrangement of the land and the ocean on the earth's surface. But this is not true. The entire continents occupy the space on landmass in certain geometrical pattern and the arrangement is tetragonal. Land and ocean both contribute to enhance the world's economy and are rich in natural resources. However, there are various important differences in the continents and the oceans, which need to be recognized:

1. The oceans form one continuum, and the continents are separate masses.

2. The oceans have ridges located in the middle and the continents have elevated Mountains particularly along the coast.

3. The oceans have deep trenches whereas continents don't have trenches so deep.

4. Volcanoes and lava-flows are more widespread in ocean floors than on the Continents.

5. Photosynthesis (which is a basic process for green plants to manufacture food) takes place more in extent in the ocean than on the land surface. This, in recent years, has created greater interest in the ocean's resources for food-supply.

Land has always been different in appearance to different persons. To a young child, it is a playground; to a poet, an attractive topic; to a patriot, it is a symbol; to the farmers, a direct means of their livelihood; and to an architect, it is a space to build houses on. Besides, land may have deposits of iron ore, as in Southern India; coal, as in the states of Jharkhand and Assam (India); and oil, as in the Middle East countries of the world as well as in Texas and Arkansas (USA). Hence land in its complete sense is a natural asset where opportunities exist independently for all mankind.

The vast amount of landmass, which lies around us, is one of the basic resources for the rapidly growing human needs. With vast scientific knowledge, we have been able to replace various commodities for higher efficiency, but we have never been successful in replacing the needs of our land. Total replacement of land will be impossible even in the future. Hence the knowledge of land mass is quite essential, so that it may be handled adequately. In recent years, this has been very essential, where population has been increasing at a rapid rate in the world and especially in the South East Asia. Even with almost stable population growth in USA, the values of our land keep on rising. This indicates the importance of our land resources, which certainly dictates the sincere appreciation for land.

When properly used, the land may contribute to our welfare at the present as well as in the future more effectively than in the past. This needs adequate planning about its best use and the know-how, to achieve it. However, this is only possible if we fully understand 'Soil', which the land contains, because the quantity and quality of soil contained by the land mass determine the value of land, based on its location and possible use. A full knowledge and appreciation of this natural resource is only possible when we realize its nature, which is vulnerable to destruction but also suitable for constructive enhancements. Soil, being a non-renewable resource, needs to be well conserved, always and at all costs.

Previously, soil was considered as a medium for plant growth. This is still true but not entirely. As we are aware that a little depth into the earth's surface is not the only way for growing crops. Basically, soil is important for all of us, including researchers, agronomists, engineers, landscape architects, farmers, and laymen. All of them in their life-time use soil so often that they even neglect to learn about it. This negligence due to carelessness or lack of knowledge in the past has resulted in serious misuse of soils. Examples in history are there to prove that the ruins of Kish at Mesopotamia, the hanging garden of Babylon, the Valley of Nile in Egypt, the highlands of Israel, and the place of milk and honey of Jordan could not last long due to wasteful use of soil and water. Lest we recur the same blunders, we have to know more and more about soils. Particularly in Asia, the higher rate of population-growth has created excessive pressure on soil and it demands our maximum attention for its care. Protect it and it will protect you. But before initiating the protection of our soil resources, we need to be familiar with it.

SOIL

Physically, soil is a portion of the landscape, which is three dimensional, and it occupies space. Soil in its' lower depth is surrounded by no soil; in its top surface it is surrounded by atmosphere; and all around it is surrounded by other soil or soils. This three dimensional concept was first developed by a Russian scientist named Vasilli Baiselevich Dokuchaev. The same concept at the same time in late 1800 also developed in USA and the scientist to be credited for this is the late Eugene Wedlemer Hilgard.

Let us examine soil as three dimensional, which represents a soil body.

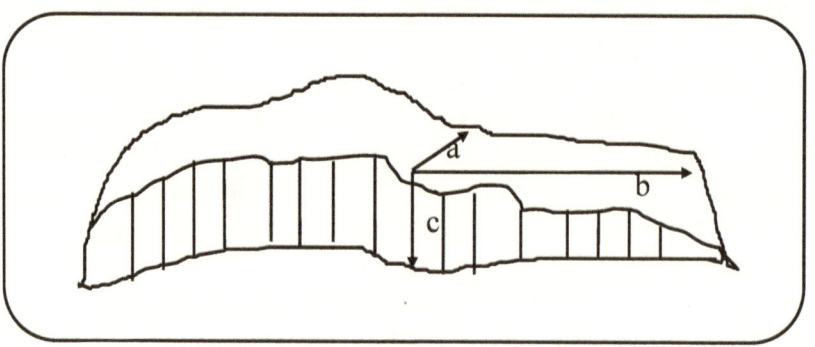

Fig. 1-1: Soil body showing in general the length a, the breadth b, and the depth c.

SOIL PROFILE

Theoretically, it may be possible to have only one soil body on a particular landscape at a particular location. But generally various adjacent pieces of soil-bodies with other accessories, like forest and human habitat combined to form a landscape. Thus when we look at a landscape we always do see either a single 'a' and a single 'b' or various 'a' and various 'b'. But unless we examine the depth 'c', the adjoining soil or soils cannot be delineated. Hence the 'c', which is a vertical deep cut in the earth's surface, exposing the soil-layers, need to be examined. This vertical deep cut in the earth's surface is known as "Soil Profile;" which is two dimensional with only b and c combined.

Soil profile may consist of one or more layers lying parallel to the land surface. These layers can be said to be horizons, when they differ genetically in one or more properties such as texture, color, structure, porosity, reaction and consistency. Besides, these horizons being genetical layers must show their individual genesis as influenced by environmental and biotic factors in course of time.

In general, a soil profile contains four master horizons-scientifically identified by the capital letters A, E, B, and C. Some soils may not have B-horizons and such soil profiles are known as AC profiles. The A horizon, the upper layer in the soil profile is often known as topsoil. Plant roots, macro, and microorganism are more active in this part, and therefore, the organic matter in this part is comparatively abundant. It is the same A horizon, which directly receives the falling drops of rain which may carry the finer soil particles to move down into the lower horizon. This makes the B-horizon richer in finer soil particles, which we find in the subsoil. Below this B, is C-horizon, which is known as "Parent Material", from which soil horizon A is derived. Later, due to leaching and environmental interactions, E and B develop between A and C-horizons. Hence it derives its properties both from A and C. The C horizons may be the deepest of the three major horizons from which true soil (with A, E, and B horizons) develops. This true soil with A, E, and B-horizons is also identified as solum and all the horizons A, E, B, when combined with C is known as regolith. On the upper surface of A-horizon, the fresh and decomposed organic matter may accumulate.

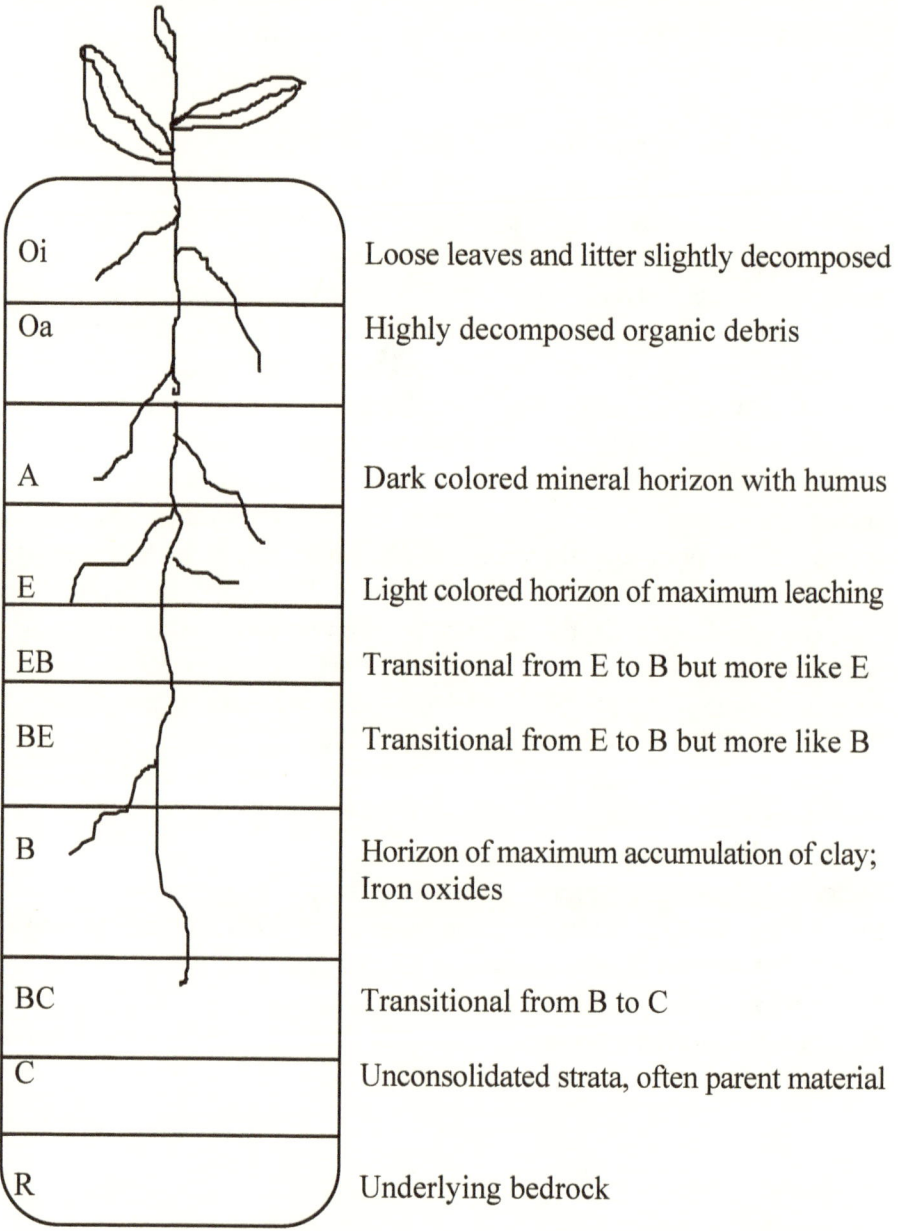

Oi	Loose leaves and litter slightly decomposed
Oa	Highly decomposed organic debris
A	Dark colored mineral horizon with humus
E	Light colored horizon of maximum leaching
EB	Transitional from E to B but more like E
BE	Transitional from E to B but more like B
B	Horizon of maximum accumulation of clay; Iron oxides
BC	Transitional from B to C
C	Unconsolidated strata, often parent material
R	Underlying bedrock

Fig. 1-2: Hypothetical soil profile in situ with all possible horizons.

However as we see the profile, the solum consists of A, E, EB,

BE, and B's where EB and BE are the part of E and B of solum having more influence of E and B respectively but at present are in transition.

SOIL COMPONENTS

When we talk about soils, we generally mean soils, which do have four major components: (1) Minerals (2) Organic Substances (3) Air and (4) Water. On volume basis, mineral matters, organic matters, air, and water are generally present in the mineral soil in the ratio of 9:1:5:5 respectively. Variation in the stated proportion is always possible in the dynamic soil system. As such water and air ratios always fluctuate and never show a constant proportion in nature. In the same token, the proportion of mineral contents and the organic contents also change. This variation may cause the soil either to be a mineral soil (having more than 80% of mineral matter) or to be an organic soil (having more than 20% of organic matter).

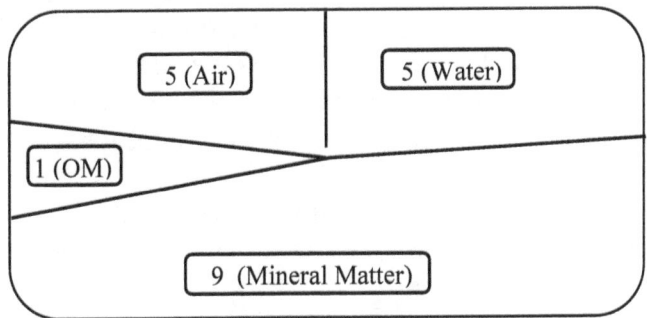

Fig. 1-3: Four components of soils in general.

These components are mixed so intimately in the entire soil that at times it is hard to separate them. These components, if they were separated, the soil's properties will be lost completely, as soil will loose its identity as a dynamic living body in nature.

SOIL FORMATION

Soil displays various qualities. Most of these qualities like porosity, drainage capacity, moisture holding capacity, plasticity, nutrient retention, rigidity, root penetration and charges on the soil particles may directly depend on its' evaluation and the stage at which the soil is being examined.

Soil, which we find at any location, is an outcome of five major factors of soil formation. Hence, Jenny (1941) very correctly interpreted soil to be the function of climate, living organisms, parent material, relief and time by representing it as a following equation:

Soil = f (climate, living organisms, parent material, relief, and time)

Parent material is the weathering of rocks, which later causes the formation of soil. This parent material may also move to other locations and give rise to soils. The former soil formation is identified as soil in situ and the later formation of soil is identified as soil not in situ. Climate and living organisms are the active factors of soil formation and the rest three i.e. parent material, relief, and time are the inactive factors.

At certain rare instances soil mass may be displaced or buried and may change with time to become parent material for other soil or may be transformed into rock.

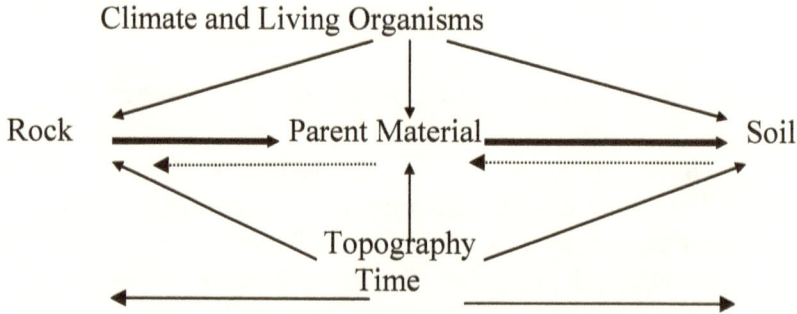

Fig. 1-4: Shows the formation of soil due to combined factors (Climate, Parent material, Living organisms, Time and Topography.)

The parent material (C horizon) may remain just over the bedrock from which it has been derived and as such the soil formation is said to be in situ.

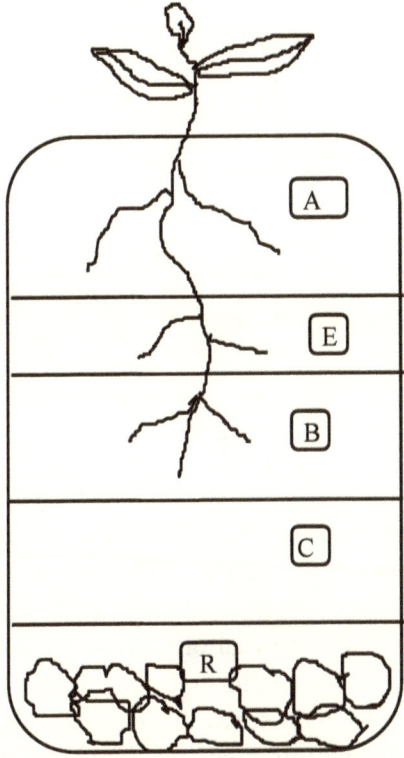

Fig 1-5: Soil profile showing A, E, B, and C-horizons as well as the bedrock from which soil has been derived.

Besides, the soil may also develop from the parent material, which has been displaced from the point of origin. In this case, the parent material may not have any relationship with the rocks beneath. Such types of formation are usually seen in Loess deposits, and in alluvial soils.

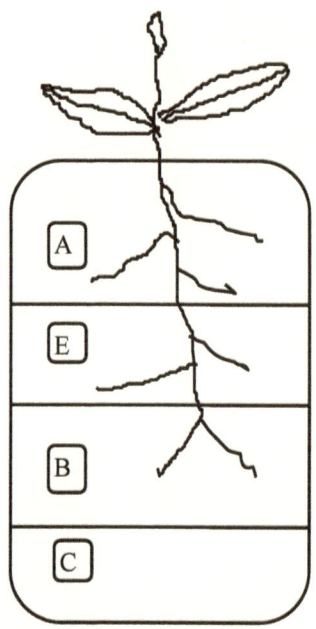

Fig 1-6: Soil profile showing A, E, B, and C horizons developed on a displaced parent material.

This mode of soil formation gives a clear-cut indication of innumerable possibilities of soils with innumerable gradations when we consider the various intensities of these five factors, which play a key role in the soil formation. Hence, scientifically speaking, soil is a collection of natural bodies on the earth's surface which have been derived from the parent material interacting with climate, living organism, as controlled by relief over periods of time. Soil as such is a natural production on the earth, which presents before us an on-going process of change. Besides, any change brought by human factors can also be accommodated in this dynamic formation of soil. Gradually, the rocks give rise to parent material, and up to that stage, there is no soil. Hence in stage one, rocks give rise to parent materials and in the second stage the parent materials give rise to soil when A-horizon is formed from the C-horizon (parent material). By that time there must have accumulated enough organic materials on the top of A-horizon to be identified as Oi or Oa. In certain circumstance, this organic layer may not be present due to locational disturbances. As times progress, the interaction of A and C simultaneously start giving rise to E and B. Therefore, E and specially B-horizons are the indications of certain advanced stages of soil formation. These varying stages of change become very noticeable when at time one soil order transforms towards the other soil order. There is a continuation of this on-going process, which continues all the time in the natural soil forming system.

BIBLIOGRAPHY

1) Anatomy of the Earth. (1968), World University Library. McGraw Hill Book Company, New York.

2) Dale T. and V.G. Carter, (1955). Topsoil and Civilization, University of Oklahoma Press, Norman. Fascinating Historical Review of Man's Impact on the soil.

3) Fitz Patrick, E.A., 1971. Pedology, A systematic Approach to Soil Science. p. 12-46. Factors of Soil Formation.

4) Foth, Henry D. and John W. Schafer. 1980. Pages 1-35. Soil Geography and Land Use. Soil Classification.

5) Hyams E. 1952. Soil and Civilization. Thames and Hudson, London. Intriguing Account of the effects of soil destruction upon history, with examples from the classical world, Far East and USA.

6) Jenny, H., 1941, Factors of soil formation, McGraw Hill, New York.

7) Joffe, Jacob S. 1948. Pedology Second Edition. Soil Genesis, pages 2-165.

8) Miller, Fred P., Wayne D. Rausmussen and L. Donald Meyer. 1985. Soil, its importance to the sustenance of human kind and historical perspective. p 23-24. Soil Erosion and Crop Productivity.

9) Opportunities In Basic Soil Science Research. 1992. Soil Science Society of America, Inc. Madison, Wisconsin, USA.

10) Planning the Uses and Management of Land. 1979. American Society of Agronomy, Crop Science Society of America and Soil Science Society of America, Inc. Madison, Wisconsin USA.

CHAPTER 2

Physical Properties of Soil

The soil formation as discussed earlier shows clear indications of the interaction between the five soil forming factors. If we add to these, the varying intensities of each of these soils forming factors, it will account for the numerous interactions resulting into innumerable kinds of soil with distinctly different properties. On similar basis, it is evident that the different soil formations may originate from the same kind of parent material. However, different parent materials may also give rise to soils of different or similar properties depending on the conditions of interactions. As a result, the soils, which we see around also, show variation in their physical properties. These physical properties can be sensed through our physical senses including touch, observations, and other sensory means. We need to understand that these properties have evolved with time. Careful account of these properties may help to read their past as well as be helpful in projecting their better use in future.

The following physical properties are discussed under the following heads: (1) Texture (2) Structure (3) Color (4) Consistency (5) Aggregation (6) Density (7) Tilth and (8) Temperature.

SOIL TEXTURE

This deals with the relative size of the individual soil separates (sand, silt and clay) based on their known ratios. Generally, the mineral soil texture falls into three textural divisions consisting of five different subdivisions, which include various textural names.

As shown in Table 2-1, the textural divisions (Fine, Medium, and Coarse) are based on the feeling that one experiences while rubbing a small amount of soil between his or her two fingers. One may find it coarse when the proportion of sand is very high and one feels it smooth when the proportion of clay is moderately high. These textural divisions may also be estimated on the basis of the resistance put by soil while plowing and are named as (heavy), (medium), and (light) as shown above. However, it is more convenient, quick, and practical to estimate the textural names of soils by rubbing between two fingers. Moisten the soil to a consistency of a workable putty. Make a ball of about 1/2-inch in diameter. Hold the ball between the thumb and the forefinger. Press your thumb forward forming the soil into a ribbon.

With fine textured soils like clay or silty clay the ribbon remains long and flexible. Soils in this group are very sticky and plastic when wet. If a ribbon is not formed, the sample is probably a silt loam or loam, sandy loam or sand; which would place it in one of the three groups i.e. medium textured, coarse textured, or very coarse textured. The decision will rest on whether sand or silt predominates in the soil sample. If the soil feels smooth and talc-like, with no grittiness, silt predominates and the soil is termed medium textured. If the soil feels slightly gritty yet fairly smooth and talc-like, it is probably a loam, or silt loam and is also included in the medium textured group.

Table 2-1 : Soil textural groups

Fine or Heavy		Medium	Coarse or Light

Fine Textured	Moderately Fine Textured	Medium Textured

Clay Silty Clay Sandy Clay	Silty Clay Loam Clay Loam Sand Clay Loam	Loam Silt Loam Very Fine Sand Loam

Coarse Textured Loamy Sand Medium Sand Fine Sand	Very Coarse Textured Gravelly Sand Coarse Sand

A marked gritty feeling and a lack of smoothness indicate that, most likely, the sand portion predominates. The soil is then considered a coarse textured. If the soil is composed of almost separate gritty materials, it is sand and is considered very coarse textured. Soil composed of much gravel with very little fine material also falls into this classification. If the gravels present are 20 to 50% on dry weight basis, the soil is gravelly and if the gravels exceed fifty percent, the soil is known as very gravelly. A soil can possibly range from gravelly clay to gravelly sand in texture.

By now, it is understood that the mineral soil's texture depends on the mixed proportion of sand, silt and clay which can be identified by rubbing the soil samples within two fingers. But for their correct identification, mechanical analysis by hydrometric method in the laboratory is the usual procedure.

Soil, as we know, has four components (air, water, mineral matter, and organic matter) and it belongs to three-phase (solid, liquid, and gas) system. In dealing with its mechanical composition of sand, silt and clay; one is concerned with only the solid phase with organic portion excluded. In determining the soil texture in the laboratory by hydrometer method three basic principles need to be observed: (1) organic matter to be discarded by mild chemical treatment without destroying the mineral portion, (2) Sand, silt and clay fractions, which may aggregate, are broken by mechanical agitation, and (3) Silt and clay particles are kept separated in solution simply by agitation and by sodium (Na^+) ion introduction at a very very diluted rate. Na^+ ion takes the water of hydration around it and separates the particles. The other alternative is to use the surface-active agents like Calgon. Finally, the percentage of total sand is determined by sieving. The silt and clay fractions are determined by hydrometric reading. (The reader is advised to consult the standard laboratory procedure for mechanical analysis). However, it is imperative that the material larger than 2mm is called pebbles or stones, and soil material only below 2mm in diameter belongs to soil. Only this fine earth is normally considered in soil analysis (mechanical, chemical, biological, or mineralogical). The boundary between silt and clay has generally been set at 2 microns (.002 mm) in diameter.

The boundary between sand and silt is taken at 0.02 mm to 0.05 mm in diameter based on the particular system used. Basically there are two classification systems as noted in Table 2-2.

Table 2-2: Various size limits of soil separated according to two systems of classification.

Soil Separates	Diameter Limits of Separates	
	USDA System, mm	International System, mm
Very Coarse Sand	2.00 - 1.00	
Coarse Sand	1.00 - 0.50	2.00 - 0.20
Medium Sand	0.50 - 0.25	
Fine Sand	0.25 - 0.10	0.20 - 0.02
Very Fine Sand	0.10 - 0.05	
Silt	0.05 - 0.002	0.02 - 0.002
Clay	Below 0.002	Below 0.002

However realizing the importance of silt and clay fractions of soil (they being more viable and reactive than the sand fractions), further fractionation of silt and clay has become an important tool in research. Therefore silt has been grouped into three sizes and clay in two as shown in Table 2-3.

Table 2-3: Diameter of the silt and clay fractions

Fractions	Diameter (mm)
	Ranges From
Silt	0.05 to 0.002
Coarse Silt	.05 - 0.02
Medium Silt	.02 - 0.01
Fine Silt	.01 - 0.002
Clay	Less than 0.002
Coarse Clay	.002 - 0.0002
Fine Clay	Less than 0.0002

It is very important to identify the sizes of these fractions, because as their sizes become smaller, they may have immense increase in their surface areas, with no change in the total volume. Figure 2-1 precisely illustrates this phenomenon.

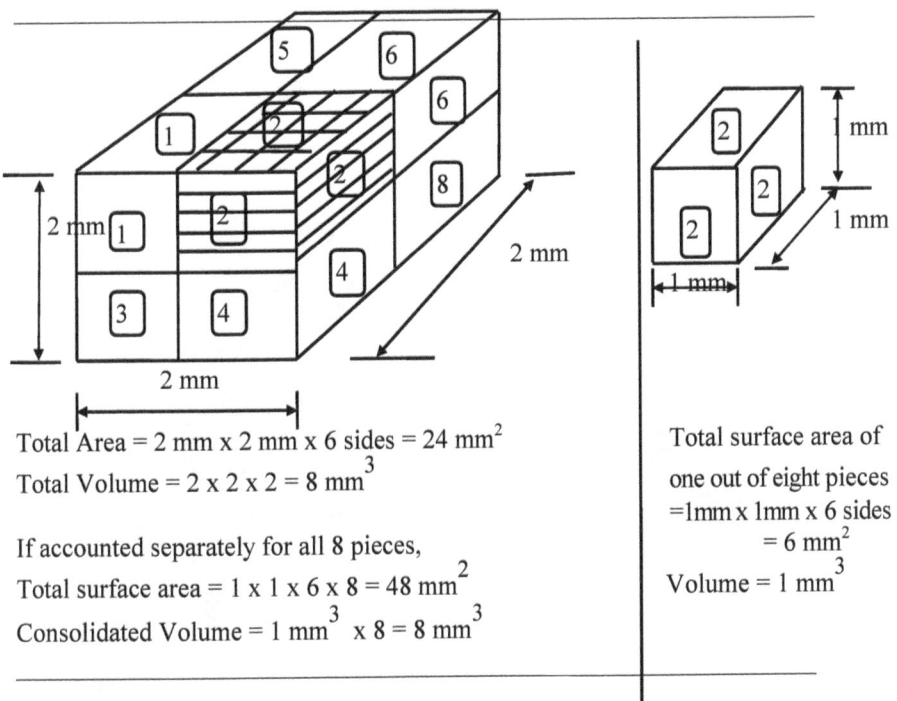

Total Area = 2 mm x 2 mm x 6 sides = 24 mm^2

Total Volume = 2 x 2 x 2 = 8 mm^3

If accounted separately for all 8 pieces,

Total surface area = 1 x 1 x 6 x 8 = 48 mm^2

Consolidated Volume = 1 mm^3 x 8 = 8 mm^3

Total surface area of
one out of eight pieces
=1mm x 1mm x 6 sides
= 6 mm^2

Volume = 1 mm^3

Fig. 2-1 : Increased surface area based on decreasing particle size.

Clay particles, which are less than .002 mm in diameter, generally gain enormous amount of surface area. This, on the whole, increases the reactivity of the soil with the natural liquid and gaseous phases with which it naturally interacts.

When the total portions of sand, silt, and clay fractions are known, the textural name of that particular soil can be given with the help of the textural triangle of soil shown in Fig. 2-2.

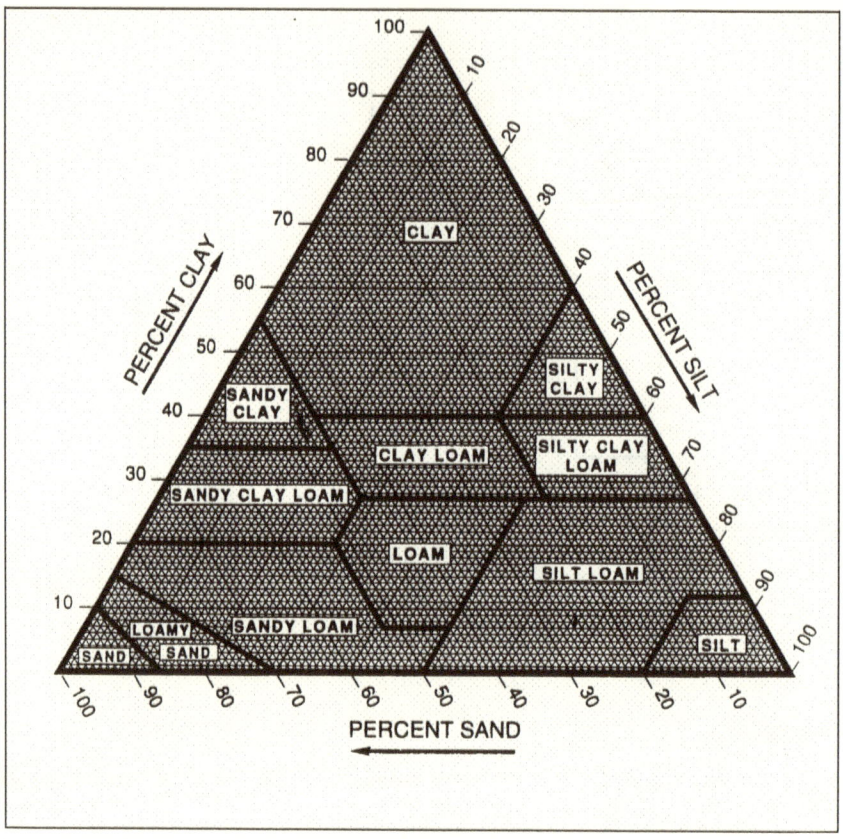

Fig. 2-2 : The textural triangle (USDA, Soil Conservation Service)

Assuming that a soil sample contains 20% clay, 40% silt, and 40% sand, the textural name of the soil will be loam. How? Place your finger on clay line at the point of 20% and move parallel to sand line within the triangle. Similarly locate the 40% point on the silt line and move parallel to clay line; these lines will intersect at a point where the third line will also intersect provided you take 40% point on sand line and move parallel to silt line. The point of intersection lies within the area of loam. Thus the soil texture is loam.

It is worth noting that to qualify as sandy soil, sand must be more than 85%; to qualify as silty soil, silt must be more than 80%; but to qualify as clay soil, the percent of clay need to be only 35% on dry weight basis. This indicates the greater effectiveness of clay particles compared to sand and silt fractions. On equal weight basis clay has large surface area even with small volume and as such it is more effective as compared to sand fractions. Sand, as a matter of fact, serves more as a framework around which silt and clay (the so-called active and potential portion of the soil) are associated. However, sand facilitates the drainage of excess water and helps the circulation of air at a rapid rate, whereas silt and clay act like the reservoir of available water and plant nutrients. A farmer, a landowner, or a country, which intends to develop its resources, should start with the conservation of silt and clay. These potential constituents of soils are too precious to loose. However, a gardener prefers a medium to light textural soils and at times adds sand to his garden area. This improves the tilth of his soil, which becomes easy to handle. This also allows water and air to permeate freely in the rhizosphere.

SOIL STRUCTURE

Soil structure refers to the aggregation of individual soil particles into large units, oriented in a manner, which gives a geometrical shape or form. Table 2-4 identifies the possible soil structures along with their geometrical orientations in terms of a, b, and c axes. Note that a and b are the horizontal axes, c is a vertical axis; all of which are imagined to pass through the center of the three dimensional soil aggregate oriented to represent its natural and original position in the soil system (Figure 2-3).

Table 2-4 : Structures of the soil system
Types of soil structure

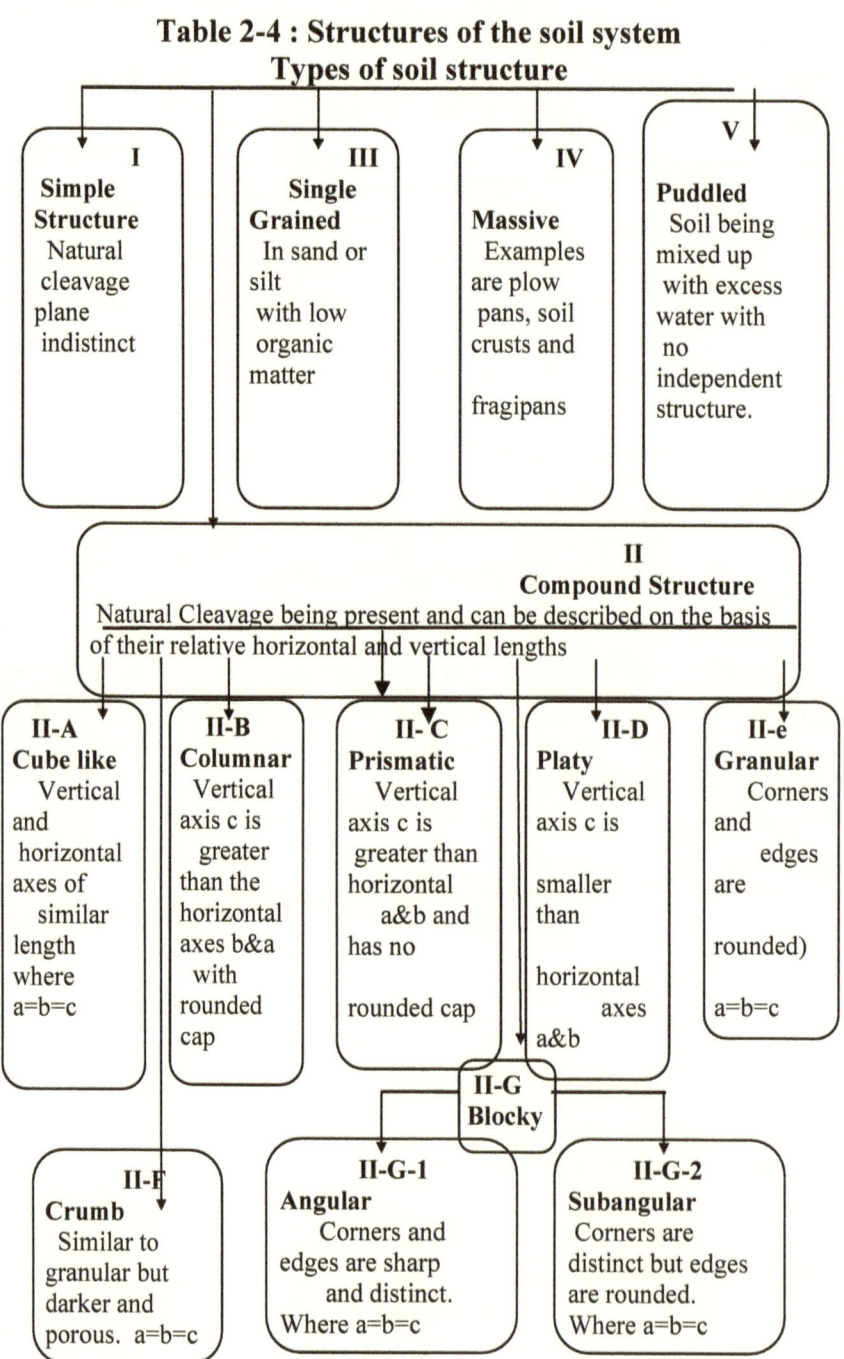

I
Simple Structure
Natural cleavage plane indistinct

III
Single Grained
In sand or silt with low organic matter

IV
Massive
Examples are plow pans, soil crusts and

fragipans

V
Puddled
Soil being mixed up with excess water with no independent structure.

II
Compound Structure
Natural Cleavage being present and can be described on the basis of their relative horizontal and vertical lengths

II-A
Cube like
Vertical and horizontal axes of similar length where a=b=c

II-B
Columnar
Vertical axis c is greater than the horizontal axes b&a with rounded cap

II-C
Prismatic
Vertical axis c is greater than horizontal a&b and has no

rounded cap

II-D
Platy
Vertical axis c is

smaller than

horizontal axes a&b

II-e
Granular
Corners and edges are rounded)

a=b=c

II-G
Blocky

II-F
Crumb
Similar to granular but darker and porous. a=b=c

II-G-1
Angular
Corners and edges are sharp and distinct. Where a=b=c

II-G-2
Subangular
Corners are distinct but edges are rounded. Where a=b=c

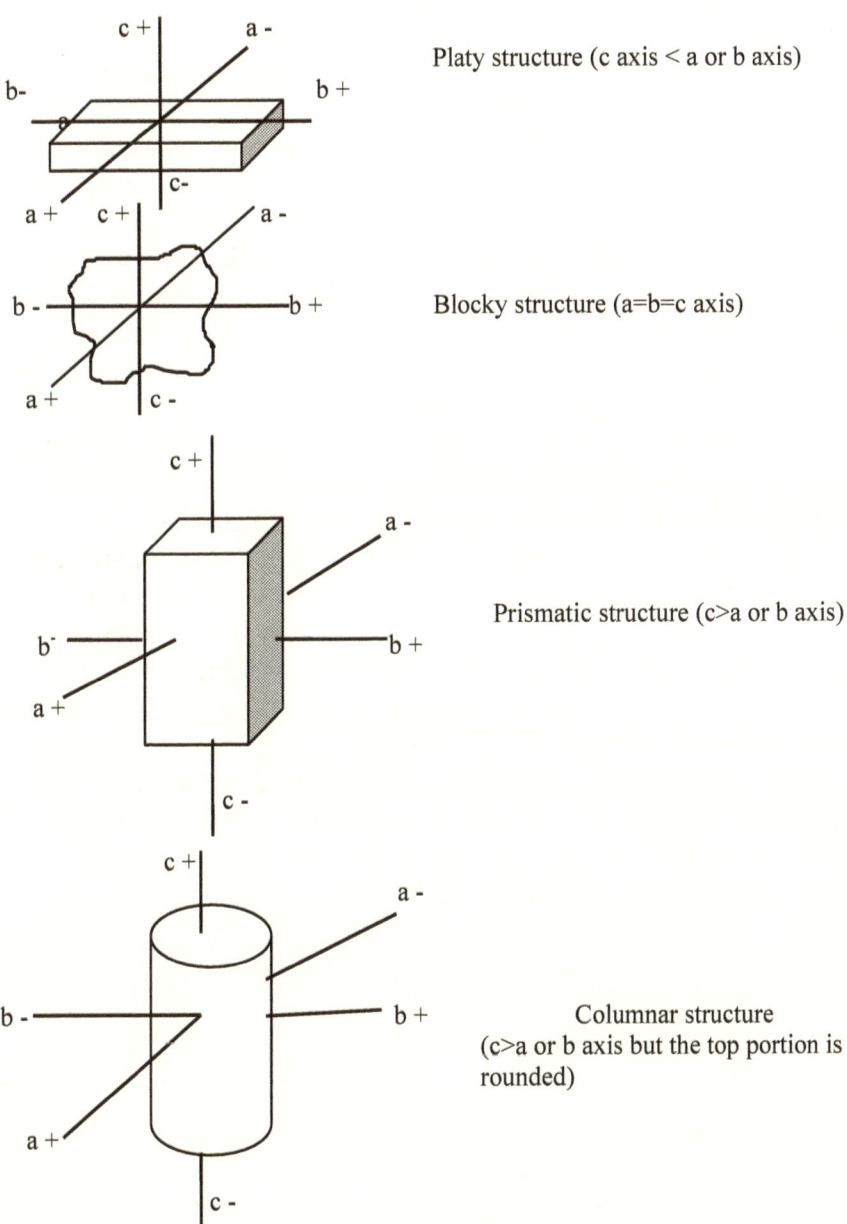

Fig. 2-3 : Geometrical orientation of platy, blocky, prismatic and columnar structures showing the relationships of a, b, and c axis.

Suresh Chandra Tiwari

These soil structures are developed and found in various environmental locations. However, certain locations are more closely associated and favorable for certain soil structures. Such relationship of soil structure to its location may be a great practical help in identifying the various structures of soil, which exist in nature. Single grained structures will be found on sandy deserts, seashores or on the dry riverbeds. Massive structures are possible in parent materials, or by continuous plowing at the same depth. Puddle soils may be easily associated with the rice fields in Southeast Asia or in Mississippi delta, which have been just prepared for seedling transplantation. Granular structures are found on surface soil or in A- horizon. Crumb structures are comparatively small, porous, and weakly held together and can be observed on surface soil with higher content of organic matter as compared to granular soil structure. Blocky and prismatic structures are found in sub soils or in B-horizons. Platy structures are found in riverbanks as well as in A, B, or C-horizons of silty textural soils. Columnar structures are found in B-horizons with dominance of Na+ ions with higher salinity. Climate plays an immense role in influencing certain environment. Therefore, wetting and drying cycle, or freezing and thawing do always play important role in formation of soil structure and its stabilization. In describing soil structure for identifying as well as for its proper manipulation and handling, it is essential to identify the structural class (which refers to the size of the ped) along with their types. Table 2-5 will help in placing the soil peds in various classes of soil structures.

Table 2 -5: Structural types and structural class

Type	Class				
	Very Fine	Fine	Medium	Coarse	Very Coarse
Platy	1mm	1-2 mm	2-5 mm	5-10 mm	10 mm
Prismatic Or Columnar	10 mm	10-20 mm	20-50 mm	50-100 mm	100 mm
Angular Blocky or Subangular Blocky	5 mm	5-10 mm	10-20 mm	20-50 mm	50 mm
Granular	1 mm	1-2 mm	2-5 mm	5-10 mm	10 mm
Crumb	1 mm	1-2 mm	2-5 mm	———	———

In nature, formation of soil structures is made possible due to the presence of (1) clay minerals (2) organic colloids (3) oxides of aluminum and iron (4) root's activities and (5) by microbial actions. However, in practice, these structures can be manipulated by adding organic matter and/or by plowing. Hence whatever structures we see in the land being utilized for crop production are formed as a net result of natural cycle combined with the human efforts of supplying organic matter, plowing and crop uses. Plant roots and residues do always add organic matter to the soil. However, the contribution of soil structure by green manuring with leguminous crops may be unparallel and very conducive to plant growth. Researches have already shown that monoculture of corn has detrimental effect on yield as a result of poor soil structure; whereas corns rotated with legumes have stabilized soil aggregates and the yields have been observed to be significantly higher.

Minimum tillage has shown profit in farming. This process can safely be used with the use of herbicides to minimize the numbers of disking and plowing and to cut down the useless competition with the weeds. The process of minimum tillage with high accumulation of organic residues has always improved the soil structure to an optimum standard for plant growth. Granular or Crumb structures may thus bring an indirect effect, which may be important in achieving higher economic returns.

Charged colloidal particles and water molecules initiate structures in soil system. Dipolar water molecules attach these colloidal particles to each other. In the process of this linking, the absorbed cations do also play a role in structural stability. This stability of soil structure is achieved by the intervention of natural wetting and drying cycle, which always continues. According to this theory, use of lime undoubtedly introduces calcium ions in soils. However, this direct effect of calcium ions in granulation of soil is surpassed by the increased amount of organic matter and microbial activities as a result of liming.

Structures are always influenced by soil texture. However, manipulation of soil structure cannot change soil texture, but may help in the improved circulation of soil air and soil moisture. The crumb and the granular structure are the most desirable types of structures, as they do have the maximum proportions of openings within and between the soil aggregates. Blocky structure is intermediate and platy structure is most undesirable: Platy structure hardly leaves any vertical opening and as such obstructs the infiltration of water and air which may permeate extremely slowly and result into increased surface erosion.

SOIL COLOR

It is one of the most noticeable physical characteristics of soil. There are approximately two hundred different colors, which we may find in various soils of the world. Even in an individual soil profile we notice marked variation in color in its different layers and at times within any particular layer itself. Along with other features, color in a soil profile is a very special feature, which helps a soil scientist to delineate one soil layer from the other.

Soil color is a result of light reflected from the soil, which depends on the following:

1. THE INITIAL LIGHT QUALITY AND ITS INTENSITY:

The quality, which indicates the various spectra of light, will invariably influence the soil color. Besides, it will also bring varying reflections of the soil being examined. Hence, as far as practicable, these conditions should be maintained uniform for comparing soil colors of various locations. Soil colors are usually examined under white outdoor daylight around noon. Determination of soil color within two hours of sunrise or sunset should be avoided.

2. MOISTURE CONTENT:

This will also influence the soil color. Moist soil will always have darker color than the dry soil because moist soil absorbs more and reflects less light. Hence in order to maintain uniformity, the moisture conditions of various soils being examined should be quite similar. In order to avoid any problem arising from this, the color of a soil should be reported in dry as well as in moist conditions.

3. THE MINERAL CONTENT OF THE SOIL:

A very small amount of iron minerals may develop colors in soils. In tropical and subtropical areas hematite (Fe_2O_3) is responsible for the red coloration under moderately well drained soil profiles. Presence of even small quantity of this mineral may impart red color. Goethite ($HFeO_2$) is the other mineral, which shows colors in soil ranging from reddish brown to yellowish brown colors, which are invariably caused by limonite ($Fe_2O_3.3H_2O$). Besides, under reduced conditions in wet soils, various gray, olive and blue colorations appear in the presence of iron. A pale gray coloration may be due to the presence of quartz (SiO_2), feldspars (plagioclase) and kaolinite $Al_4Si_4O_{10}(OH)_2$; whereas manganese dioxide (MnO_2) may cause dark coloration. Presence of olivine $(MgFe)_2SiO_4$ and chlorite $(Mg, Fe, Al)_6(Al, Si)_4O_{10}(OH)_8$ may cause green coloration. Muscovite $KAl_2(Al Si_3 O_{10})(OH)_2$ may appear in brighter tints.

4. SOIL REACTION:

Soil pH influences the soil colorations indirectly. In low pH soil, organic matter in presence of low calcium imparts pale coloration, whereas under high calcium or sodium with higher pH even a small amount of organic matter gives rise to dark colored soils.

5. SOIL ORGANIC MATTER:

Organic matter increase in soil has been very well correlated with increasing darkness of soil color. This is usually observed that the color of the upper top horizon in uneroded soil appears in general to be dark brown to black in color. This color fades as we go down in the soil profile with the decreasing amount of organic matter. Organic matter may appear darker with greater humification. Even in tropical conditions, the moist soil contains more organic matter as it resists against the rapid depletion and may maintain darker coloration depending upon the moisture status of soil.

Thus we find that soil color in nature obviously presents insight on any individual soil or even on soil layers within a profile, provided it is interpreted wisely. This has given rise to a real need for measuring and expressing the soil colors as accurately on scientific basis as possible.

In recent years, Munsell color charts have been developed and are used to read soil color. In the making of Munsell color charts, the three basic components of color (Hue, Value, and Chroma) have been considered. Hue refers to the dominant spectral color or quality of light, which distinguishes red from yellow as based on the wavelength. Value refers to the quantity of light ranging from black to absolute white, which in other words expresses the apparent darkness as compared to absolute whiteness. Chroma refers to the gradation of the purity of color or the measure of the degree of departure from neutral grays or white.

Each page in the Munsell color book refers to a particular hue within which various color chips are systematically arranged. These color chips are arranged vertically upwards with increasing value and being separated by a circular hole where the soil ped can be placed from below for color matching. Horizontally from left to right these color chips show increasing chroma. Hence, each of the color chips has been arranged vertically as well as horizontally. Each of these represent certain value and certain chroma with reference to the particular hue as indicated in the Munsell color book. Suppose we are looking at a page belonging to hue 10YR. On matching a soil ped the color chip that matches exactly with the ped may have value of 6 and chroma of 4. Thus, the soil ped being examined belongs to the specific color 10YR 6/4 (light yellowish brown).

The complex nature of soil indicates that we may find all colors including white, red, gray, brown, yellow and black. Even bluish and greenish colors generally occur in mixtures and may appear as brown, rusty, gray and dusty white. Under waterlogged conditions alternated with the drying effect, we do find two or three colors occurring in patches, which are known as mottling. These are caused due to reduction and oxidation of iron and manganese. Reduction under water causes them to precipitate out. The presence and the concentrations of these mottles do indicate the poor internal drainage of soils and at the same time corresponds to the concentrations of iron and manganese in soils.

Soil color has definite use in the comprehensive system of soil classification. The formative elements as "chroma" to indicate the presence of pronounced color, "ochre" (light colored), "umber", (dark) and "alb" to indicate white coloration in soils have been used. Soil color may also well be related to climatic zones. Tropical soils are generally red and yellow. Temperate and cold zone soils may be dominantly grayish. Physiographic effect on soil color has also been observed. Upland soils as compared to bottomland soils have always appeared to be lighter in color. Soils derived from the basic rocks may have colorations deeper than soils derived from acid rocks. These are some of the various effects, which we find on soil color, which reveals some important aspects of that soil. Soil description remains quite incomplete without color identification. At this age and time of energy crisis, darker soil color does a better job in absorbing solar energy. As we grow more and more knowledgeable regarding soil, we will find soil color to be of increasing importance.

CONSISTENCY

Soil's consistency describes the resistance of soil to any given mechanical stress based on its cohesive and adhesive properties. Cohesive property in soils describes the properties of soil materials attached to the adjacent soil materials directly; whereas the adhesive properties describes the attraction of soil materials to other gaseous or liquid phase in the soil system. It is certain that the cohesive and the adhesive properties of soil do change under different moisture levels in soil, which always keep on changing with time under natural environment. Hence constancy of soil needs to be described in reference to the moisture status of soil in all conditions. The soil's behavior can be wisely predicted provided the consistencies of soil under dry, moist and wet conditions have been identified.

The field criteria (dry, moist and wet conditions of soil) that eventually control the soil's consistence under the naturally set conditions in the field are noted below:

(I) Dry Consistence - (Air-Dry):
(a) Loose - falls apart without handling. A ped cannot be picked up.

(b) Soft - The soil mass can be picked up but falls apart with slight pressure.

(c) Hard - Must exert strong pressure to break. However, it can be broken between thumb and the forefinger under the strongest pressure one can exert.

(d) Very Hard - The ped cannot be broken under the strongest pressure applied between the thumb and the finger.

(II) Moist Consistence - (This is in between the air-dry and field-capacity in its' moisture content):

(a) Loose - soil ped cannot be picked up as it falls apart.

(b) Very Friable - Crushes with only slight indentation of fingers.

(c) Friable - When crushed it indents the fingers but only gentle pressure is needed.

(d) Firm - It crushes only when deliberate pressure is applied. It deeply indents the fingers.

(e) Very Firm - Can hardly be crushed between the thumb and the finger.

(f) Extremely Firm - Cannot be crushed between thumb and the forefinger.

(III) Wet Consistence (Slightly above field capacity):

(A) Stickiness - Press between the thumb and the finger to identify.

> (A-1) Nonsticky - Almost nonadhereble to either fingers.

> (A-2) Slightly sticky - Adheres to both fingers but cleanly pulls out free without stretching.

> (A-3) Sticky - Stretched noticeable before breading and leaves materials on both the fingers.

> (A-4) Very Sticky - stretches as one exerts strong effort to pull fingers apart.

(B) Plasticity (Roll and deform to identify):

> (B-1) NonPlastic - By rolling with fingers, it cannot form a wire.

> (B-2) Slightly Plastic - Can form a wire by rolling. The wire cannot support its own weight. Easily deformed under pressure.

> (B-3) Plastic - Can form a wire that can hold its own weight. Must press to deform.

> (B-4) Very Plastic - Can form a wire that will hold strongly and may need strong pressure to deform.

SOIL DENSITY

Density of any material depends on its mass as well as its volume. Mass per unit volume of the material is density, which can be represented as gm/cc if mass is recorded in gms, and volume is recorded in cc. However, in case of soil, we need to think of two kinds of densities: (1) Particle density and the (2) Bulk density, because soil as we know consists of solid materials (minerals and organic matter) as well as pore spaces (air and water). When we consider only the solid materials of soil as the mass and the total volume of this solid material to calculate the density, this will be the particle density of soil. However, if the solid material plus pore spaces are considered together as a mass of the bulk soil and the volume of the bulk soil is taken into account, we will come up with the bulk density of the soil.

The following details will clearly indicate the difference in these two densities of the same soil: A one cubic centimeter (cc) of natural soil under dry condition can be represented by a volume which is one square cm in area and one cm in depth as shown below.

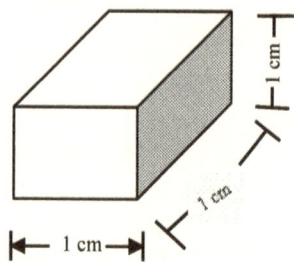

If the same soil of one cubic centimeter is compressed, the volume of the bulk soil mass decreases. Considering there were 50% pore space and 50% soil mass, the loss in volume for this one cubic centimeter soil will be 1/2 cubic centimeter. Let us compare these two cases: (1) compressed soil and (2) uncompressed soil, as shown below:

1. Compressed soil
   (for Particle Density)

2. Uncompressed soil
   (for Bulk Density)

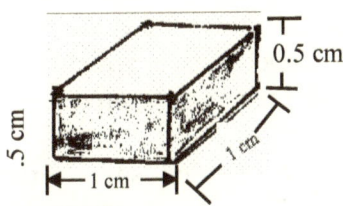

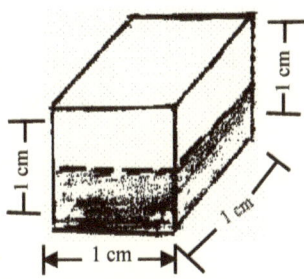

Volume = (1x1x0.5) = 0.5 cm^3
(solid soil only)

Volume = (1x1x1) = 1 cm^3
(solid + pores)

weight = 1 gram
(of solid only oven dried)

weight = 1 gram
(of oven dry soil + pores)

Hence Particle Density
= 1g/0.5 cm^3 = 2g/cm^3

Hence Bulk Density
= 1g/1 cm^3 = 1g/cm^3

It is worth noting that in both cases the weight of the soil solids, and the weight of the soil solids plus the pore space under oven dry conditions are considered to be the same, because the weight of air under oven dry conditions in uncompressed soil is significantly negligible. However the volumes of the same soil in both cases are quite different with compression and without compression.

There are definite variations in particle density of individual soils that ranges between 2.60g/cm3 to 2.76/cm3. This narrow range is based on the minerals like quarts, feldspar, micas and the colloidal silicates, which generally dominate the soil, and they do have the densities within this range. However, the variations in organic matter contents may create the difference in the particle densities of surface soils as compared to sub soils, where the particle density shows a comparatively lower value than the sub soils having less organic matter. Therefore, it needs to be clearly understood that the contents of heavier minerals like magnetite, zircon, epidote and tourmaline in soil may increase the soil density values. On the other hand, the values of soil densities may decline due to increasing contents of soil organic matters.

Bulk density generally depends on the pore space. The higher the pore spaces, the less will be the value of bulk density. Finer particles like silt and clay being porous can hardly loose their pore spaces as compared to solid sand particles. Therefore the contents of the finer particles in soils may lower the bulk density, which may range in between 1 to 1.6 in heavier textured soils as compared to 1.2 to 1.8 in light textured sandy soils.

However, the compactness in coarse textured soils varies gradually as compared to fine textured soils. This also indicates that fine textured soil may have wider range of compactness (from very low to very high) and the coarse textured soil may have narrow range of compaction (from low to medium).

Thus, there seems to be a relationship between bulk densities, texture, and compactness of soil as indicated in Figure 2-4.

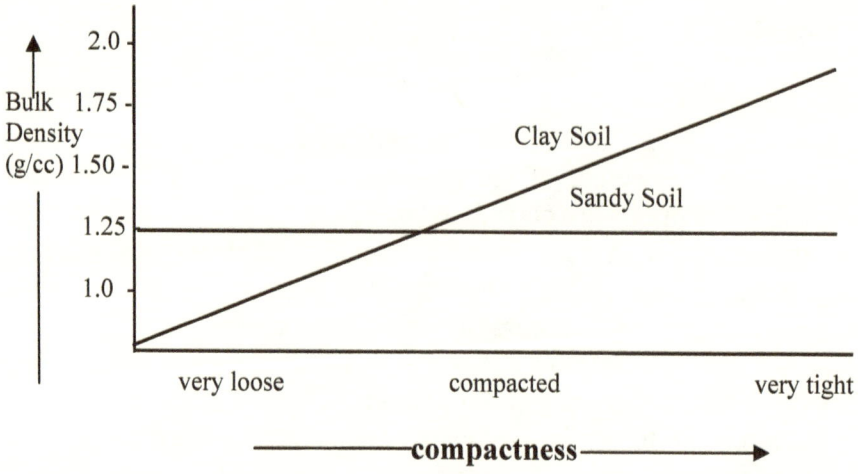

Fig. 2-4: Clay showing wider ranges of compactness than sand based on the bulk density.

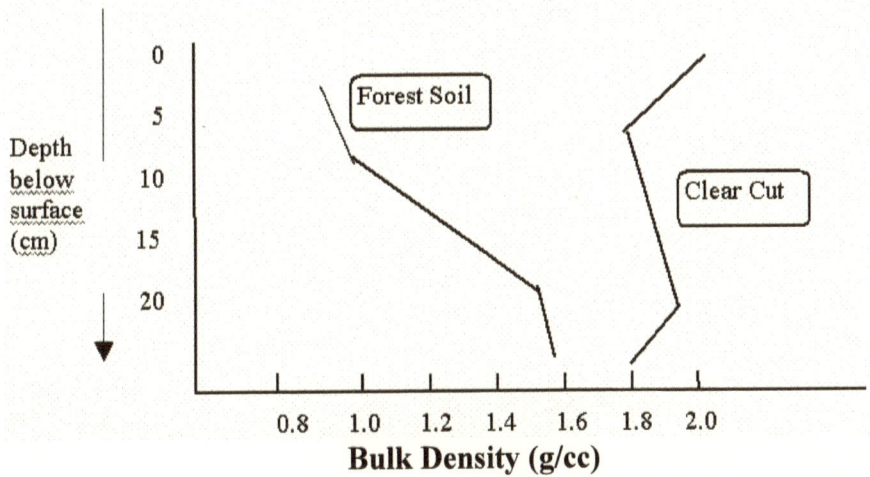

Fig. 2-5: The change of bulk density with soil depth under forest canopy and clear cut.

In spite of all the background concerning soil densities (particle and bulk), one has to have a clear understanding of the mathematical relationships that exist in soil system within various parameters (pore space, solid space, bulk, and particle densities). Each parameter is unique and each one may influence another as indicated by the following:

(1) % pore space + % solid space = 100

(2) % solid space = (Bulk Density / Particle Density) x 100

It has also been observed that soil under forest vegetation may have sharp change in the bulk density due to canopy cover influencing the yearly input of vegetations on the top soil as compared to its subsoil and/or the adjoining clear cut areas as shown in Figure 2-5.

AGGREGATION

Stable soil structure is known as aggregation. Soil aggregation is of prime importance for the farming community. It has been observed that if the % of soil aggregation exceeds 50%, the yield of crops may be well sustained at higher level. The figure that follows clearly depicts this phenomenon:

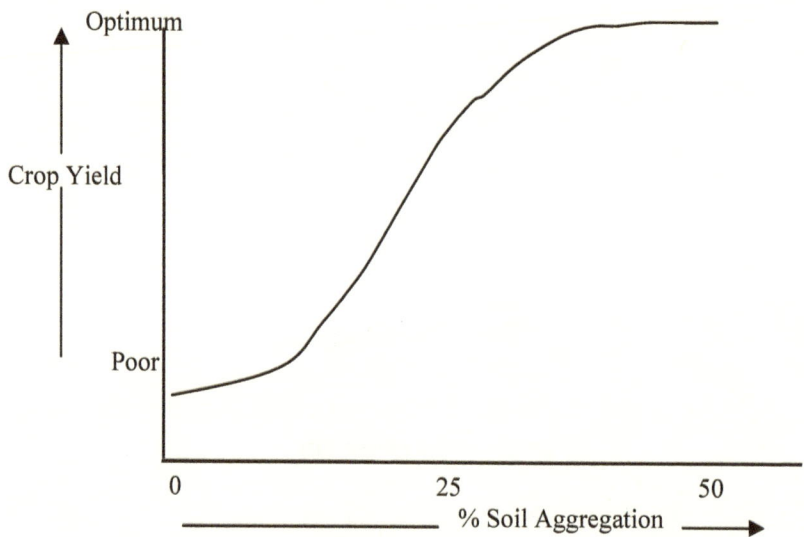

Fig. 2-6 : Percent soil aggregation as relates to crop yields

Thus the soil aggregation is an important aspect of soil management that can be formed and maintained by combination of approaches. We need to examine the following factors, which help in initiating as well as in the maintenance of soil aggregates.

(1) IMPACT OF ORGANIC MATTER IN SOIL AGGREGATION:

The organic matters are sources of life in soil system, which is mostly inhabited by bacteria, and fungi and the secretions from these organisms may bind the soil particles to give a stable soil structure. Besides, the organic secretions from the plant roots as well as various organic compounds from the plant biomass may combine with the silicates as well as the oxides of iron and aluminum which may encourage the binding actions of the soil particles to keep them together. The ever changing and dynamic material known as the end product of organic matter is 'humus' which plays an important role in stabilizing soil structure by forming organo-mineral combinations, which are so useful in farming for sustaining higher yields.

(2) PHYSICAL PROCESSES:

There are various processes, which occur in nature constantly, or intermittently, which can be listed as the physical processes. These are of prime importance that cause soil aggregation and are being noted below:

(1) Physical activities caused by roots and soil organisms.

(2) Natural cycle of wetting and drying causing swelling and shrinking of soil.

(3) Natural cycle of freezing and thawing bringing changes due to varying soil temperature.

(4) Physical manipulation of soil triggered by tillage.

Anyone or the combination of all the above processes creates movement of the soil system, which under different ecosystems may create stable or unstable soil structures, which can be evaluated quantitatively.

(3) ADSORPTION OF CATIONS:

Cations like Ca^{++}, Mg^{++}, Al^{+++}, and K^+ may encourage aggregation by their adsorbed action encouraging flocculation of the soil colloids. On the other hand, ions of Na+ may disrupt this by causing deflocculation and dispersion of the soil particles resulting into the destabilizing action against the aggregates.

(4) SOIL CONDITIONERS:

Some polymers (polydronides, and polysaccharides) do also have the ability to maintain soil structure if used as a spray on the soil surface after tilling. These organically synthesized materials may be of greater importance if they become economical in use as soil conditioners. It appears that the combined effects of these inputs (physical as well as biochemical) may be the best way to maintain stable soil structure of this viable soil system. In order to achieve this, it is imperative that certain specific minimum amount of soil organic material in soil should always be maintained, which may vary from one ecosystem to another.

TILTH

The term 'Tilth' has been defined as optimum condition of soil for plant growth. In order to attempt to reach this condition, soil may need to be tilled for seedbed preparation as well as to build stable soil structure at least up to the specified crop-growing season. This kind of soil condition has close relationship with soil moisture, bulk density, infiltration, as well as with the degree of aeration. However, on long-term basis, this approach of regular tilling may disrupt the tilth of soil due to loss of organic matter in the upper soil layer as well as by the high compressibility caused by tilling equipment in the soil zone. Hence the recent approach calls for long-term maintenance of tilth and adequate soil management by minimizing the tilling of soil and enhancing the organic materials on surface soil by plant residue management practices. We need the soil to be sufficiently loose for the root-penetration, aeration, moisture and nutrients' movement but at the same time it may be sufficiently strong to hold the plants in place. This itself is an unapproachable objective on cent percent basis. Hence we need to understand that the maintenance of tilth depends on soil mineralogy, organic matter contents, soil texture, as well as its moisture and aeration which depend on the distributions of their soil pores (micro as well as macro). But in general, it should be clear that the finer particles of soil (clays) are more compressible than the coarser particles of soil (sand). Therefore, these two extremes of soil texture are hard to manage for their tilth as compared to loamy soil. However, better understanding of these conditions may lead to better tilth-management in any soil.

SOIL TEMPERATURE

Maximum source of energy for plant growth comes directly from the sun. This energy does not only control the upper biomass of green plants by photosynthetic processes but also directly affects the rhizosphere by helping seeds to germinate followed by stable uniform distributions of many kinds of plants on the earth's surface based on varying soil temperatures on daily and seasonal basis. This is the basic factor controlling the natural spread and adaptability of various plant species on the entire planet earth. This radiant energy from the sun directly controls the planet's adaptability by affecting the physical, chemical, and biological rates within the soil system. Under similar plant growth situations, the increased soil temperature has always shown increased yields on comparative similar inputs.

Moist or wet soils warm slowly and cool slowly, whereas dry soil warms and cools in comparatively less time. Such fluctuations may have sharp occurrence in the upper soil body than in the lower soil depths. Rainfalls in different seasons in temperate regions affect the soil temperatures differently. In early spring rainwater may have warming effects on soil in the temperate regions and may have a cooling effect in the summer. Such occurrences become more obvious in tropics where rain causes the cooling effects in the summer and warming effects in the winter.

Fifty percent on average of the Sun's radiation reaches the earth surface, and only 10% of which strikes the soil system to warm the soil. The reflection, absorption and deflection of heat energy primarily depend on the Sun's angle being based at particular time and location with respect to the earth's surface. Part of the solar energy is either scattered or absorbed before reaching the earth's surface. A part of that which may touch the earth's surface is either utilized in evapotranspiration of green plants and/or in warming up the soil and its' moisture. Another very important factor is the mulching effect, which affects the solar radiation. However, in general the mulched soils do have less fluctuation of soil temperature than the unmulched or bare soil. Besides, soil moisture or mulching may have their mutual effects on each other and they do have the similar effects, if assessed separately, in controlling the soil-temperature fluctuation, which is very desirable for good seed germination and root growth.

The practical way to look into the beneficial effects of soil temperature is to acknowledge the following facts:
(1) Most of agronomic crops fail to grow normally if the soil temperature is below 40° F.

(2) The use of black polyethylene mulch on soil surface in pineapple have been found responsible for the 50% increase in growth and yield of pineapples.

(3) The use of clear plastic in Alaska has successfully increased the soil temperature, that enhances the growth of corn.

(4) Specific ranges of soil temperature may be useful for specific plant growth.

For example:
        (a) 45 to 85° F for corn.
        (b) Above 80° F for melons.
        (c) 60 to 70° F for sorghum.

(5) Uptake of nutrients and especially of 'P' is retarded in the cold soil, which may be improved by adding more 'P' or out draining the soil water.

(6) Natural cycle of cooling and heating of soil stabilizes the soil structure, particularly under cloudy conditions.

GROWING DEGREE DAYS (GDD)

Soil temperature is directly influenced by atmospheric temperature. Therefore, growing degree days of any crop indirectly depends on soil-temperature. Thus, GDD calculation may be an important aspect for various crops.

From the energy point of view, growing degree days (GDD) is calculated for specific crops based on their three requirements. There are low and high temperatures, when plants cease to grow, and the optimum temperature when the plants do grow the best. There are some estimated values for certain crops, which are as follows:

Crop	Estimated Temperature (°C)		
	Low	Optimum	High
Corn	10	32	46
Tomato	10	26	32
Fruits	4	25	36

When any day the temperature becomes lower than the specific values of low or high, we note down the estimated low or high values to calculate the GDD. Total growing degree days is calculated on daily basis throughout the growing season of the crop. Suppose on any particular day when corn plants were grown in the field, the minimum temperature is 46°C and maximum temperature is 80° C; we will adjust these values to 10 and 46^0 C respectively and GDD for that day will be:

$$\{ (46 - 10) / 2 \} - 10°C = 8$$

Adding the GDD values throughout the growing season, will give an estimated value of GDD for that crop in that area.

BIBLIOGRAPHY

1) Brady, Nyle C. 1990. Physical Properties of Mineral Soils. p 91-122. The Nature and Properties of Soils 10th Edition.

2) Donald, N. Munns, and Singer, Micheal J. 1987. Solids and Pores. p. 21-56. Soil An Introduction.

3) Foth, Henry D. 1978. Physical Properties of Soils. p. 25-62. Fundamentals of Soil Science Sixth Edition.

4) Glinsky, Jan, and Jerzy Lipiec. 1990. Soil Physical Factors Influencing Root Growth. p. 1-29. Soil Physical Conditions and Plant Roots. CRS Press, Inc., Boca Raton, Florida.

5) Kohnke, Helmut. 1966. Physical Properties of Soil. p 4-12. Soil Science Simplified.

6) Miller, Raymond W. and Donahue, Roy L. 1990. Soil Physical Properties. p. 46-81. An Introduction to Soils and Plant Growth. Sixth Edition.

7) Troeh, Frederick R., and Louis M. Thompson. 1993. Physical Properties of Soils. p. 37-64. Soils and Soil Fertility.

8) United States Department of Agriculture, Soil Conservation Service, and Soil management Support Service. 1992. Keys to Soil Taxonomy.

CHAPTER 3

Chemical Properties of Soil

Soil is made up of chemicals. The unique chemistry of soil is indispensable for human existence on the planet earth, because these chemicals, one way or the other, maintain the cyclic food chain through the plants and connect the human life constantly with this cycle. In general, the soil chemistry always deals with the various aspects of solution and solid phase chemistry. Thus, the contact area between the liquid phase and the solid phase is of immense importance in the chemistry of soil, which naturally deals with the colloids. Besides, it is also interesting to note that the ions adsorbed in the solid particles in the soil system outnumber the ions in solution, thus maintaining a very dilute concentration of the nutrient elements. This bridges the gap between soil chemistry and the flow of nutrients to plants in the soil system on a very continuous basis. Such chemical properties of soil can be stabilized and maintained within certain limits under a defined ecosystem. However, the open surrounding of natural soil has always been a very dynamic system. These important ideas can be better comprehended by understanding the chemical properties of soil under the following subheads:
1. Chemical composition of solid earth's crust
2. Soil pH
3. Soil moisture
4. Cation exchange capacity of soil
5. Availability of plant nutrients for plant growth

CHEMICAL COMPOSITION OF SOLID EARTH'S CRUST

Soil is a part of the landscape. And the entire landscape is just a crust on the planet earth. Therefore the earth's crust in reality includes the soil as well as non-soil zone. We also know, the soil is constantly being generated by no-soil area beneath, with time. Besides, in this landscape it is also not impossible to have the soil sediments gradually being transformed into the no soil zone of the landscape depending on the toposequence of the latter.

The following composition of solid crust of the planet earth, which represents the approximate percent of the chemicals present including also some toxic elements (Hg, Pb, Cd, Cu, Ni and Co), are noted below:

Oxygen - 47.7%
Silicon - 27.7%
Aluminum - 7.8%
Iron - 4.5%
Cadmium - 2.3%
Magnesium - 1.2%
Sodium - 2.5%
Potassium - 2.5%
Titanium - 0.5%
Hydrogen - 1.2%
Carbon - 0.2%
Phosphorus - 0.1%
Sulfur - 0.1%
Toxic elements - 1.0%
Others - 2.1%

In the above list "others" include Mn, Zn, Mo, B, Cl, N, V, Sn, Se.

SOIL pH

The Concept: pH is defined as log (1/H^+) ion concentration where H^+ ion concentration is expressed in moles/liter or in m moles/ml. Hence pH = log (1/H+) or it is equal to -log (H^+). At 25^0 C 1 mole of water = 18 gms; and at that temperature 1000 cc of water weighs 997 gms. Now a few problems and their answers may clarify the concept more elaborately:

PROBLEM #1

If 997 gms of water is 1000 cc at 25^0 C, this is equal to how many moles of water?

SOLUTION:

H_2O (water) 18 gms = 1 mole of water at 25^0C
Then 1 gm = 1/18 mole
Then 997 gms will be = 1/18 x 997 = 55.4 moles of water
or moles of water = (997 g of H_2O) x {1 mole of H_2O)
$$18 \text{ g of } H_2O$$
$$= 997 \text{ x } 1 \text{ mole}/18 \text{ } H_2O$$

$$= 55.4 \text{ mole of } H_2O$$

PROBLEM #2

At 25^0C temperature the product of concentration of H^+ and (OH)⁻ will be 10^{-14} moles per liter. How?

SOLUTION:

Because at this temperature, 1.0 liter of water holds 55.4 moles of water; out of which 55.3999998 moles remains unionized and the rest ionizes.

HOH = 55.3999998 unionized (mole)/lit.
(H^+) =   0.0000001 ionized (mole)/lit.
(OH^-) = 0.0000001 ionized (mole)/lit.
Total = 55.4 moles of water
Hence (H^+) x (OH^-) = 0.0000001 x 0.0000001 = 10^{-14} mol/l

PROBLEM #3

Explain mole, one mole weight, and formula weight in relation to the Avogadro number.
Solution: Table 3-1 will provide an explanation.

Table 3-1. Showing the relationships between chemical name, chemical symbol, atomic weight, weight of 1 mole, Avagadro's number and kind of particles in 1 mole.

Name	Chemical Symbol	Atomic Weight (awu)	Weight of one mole (g)	Avagadro's number and kind of particles in one mole
Oxygen	O	16 awu	16 g	6×10^{23} atoms
Oxygen	O2	32 awu	32 g	6×10^{23} molecules
Sodium ion	Na^+	23 awu	23 g	6×10^{23} ion
Sodium Chloride	NaCl	58 awu	58 g	6×10^{23} ions of Na^+ ion + 6×10^{23} ions of Cl^-
Sodium sulfate	Na2 SO 4	142 awu	142 g	12×10^{23} Na^+ ion + 6×10^{23} SO_4^{-2} ions

PROBLEM #4

If active H^+ concentration $= 10^{-6}$ moles/lit, what is the pH?

SOLUTION:

$pH = \log 1/H^+$ concentration $= \log 1/10^{-6} = 6$

PROBLEM #5

In the problem above, how much OH^- will be present/liter at 25^0C.

SOLUTION:

Concentration of $(H)^+$ x concentration of $(OH)^- = 10^{-14}$/liter at 25^0C.

Concentration of $OH^- = \dfrac{10^{-14} \text{ mol/l}}{10^{-6} \text{mol/l}}$

$= 10^{-8}$ mole/lit.

PROBLEM #6

What will be the pH of that water having active $\{H^+\}$ = 0.0000001 mole/liter as shown above?

SOLUTION:

$pH = \log 1/H^+ = \log 1/.0000001$
$= \log 10000000 = 7$

PROBLEM #7

Estimate the pH of a solution if H^+ concentration = 0.003 moles/liter. This may be pH = 3, > 3 or < 3.

SOLUTION:

Active H^+ of .001 = pH 3.
Active H^+ of .01 = pH 2.
Hence, H^+ of 0.003 moles/liter is between pH 2 and pH 3, which will be < 3.

PROBLEM #8

What is the pH of the soil solution which has active H^+ = 0.000001 mole/lit.?

SOLUTION:

$pH = \log (1/.000001) = \log 1000000 = 6$

PROBLEM #9

What is the H^+ concentration of a soil solution with a pH of 3?
Solution: $pH = 3 = \log (1/.001) = \log 1000$
Hence H^+ concentration $= (1/.001) = 1/10^3 = 10^{-3}$ moles/lit.

Soil pH has a great importance for the growth of plants, which needs to be maintained at proper level. Most crops thrive best in slightly acid soil (pH 6.5 to 6.8) but generally the legumes prefer neutral soil (pH 7.0). But there are some plants like blueberries, azalea and camellia, which thrive better under definite acid conditions (pH 4 to 5.5). Therefore, monitoring pH time to time is always advised to match the suitable pH for the crop to be raised.

SOIL MOISTURE

Soil moisture can be scientifically defined as the amount of moisture contained in the soil being calculated on dry weight basis at any particular time frame. For example 30% soil moisture indicates that 130 gms of the soil collected from the field had 30 gms of moisture and 100 gms of dry soil. This soil moisture always keeps on fluctuating based on its dynamic nature in relation to its ecosystem.

Soil moisture is a very important factor for determining the agricultural production as the entire agricultural activities just oscillate in between water too much or too little. Basically, the importance of water is based on the following two things:

1.   The need of water which is so essential for the entire growth and physiologic processes to be carried on within the plant system, and

2.   The soil available moisture directly controls the nutrient's uptake in plants.

However, the total amount of water may not be as important as the amount of available forms of water for plants, which maybe called as "available water holding capacity of the soil." Physiologically, this availability of water becomes most critical at the time of pollination and fruit setting; specially, when we aim for higher yields with increased input of nutrients. Thus the water-holding capacity aspect of the soil may not change drastically but it can be improved by adequate management and regulated supply as needed. When critically examined, soil moisture is essential for (1) nutrient carrying as well as (2) for maintaining turgidity in plants. Besides, the moisture in plant system is needed for (1) Photosynthesis as well as for (2) the movement and translocation of the protoplasm.

SOIL MOISTURE CLASSIFICATION

The soil moisture has been classified by two ways: (1) Physical classification and (2) Biological classification. Both the systems of classification will be illustrated below as they relate to each other based on the chemical behavior of soil water (structure, polarity, cohesive and adhesive forces, as well as by their surface tensions).

WATER'S STRUCTURE

The asymmetrical structure of water is such that, it has positive charge on one side and negative charge on the other.

H_2O molecule =

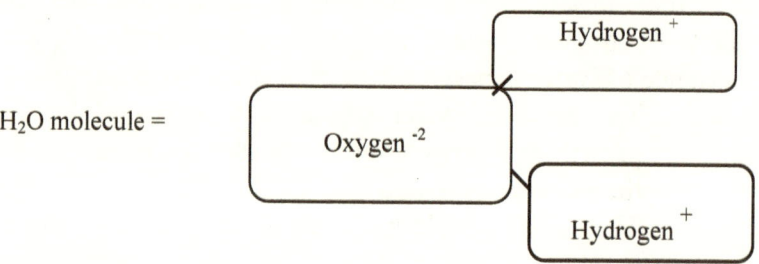

This structure of water shows the "polymer like" behavior of water molecules towards each other due to hydrogen bonding. Besides, the water molecules also get attracted to the cations held in the soil system indicating their attraction towards the negatively charged clay particles also.

COHESIVE AND ADHESIVE FORCES

Thus the soil moisture due to its polarity always shows cohesive force (one water molecule as attached to the other water molecule) as well as adhesive force (water molecule tends to attach with the soil mass).

SURFACE TENSION

Due to the cohesive force, water has higher surface tension than other liquids. This generates the capillary movement of water and indicates the amount of water, which may be retained by soil. With the above background, it will be easier to comprehend the two soil water classification systems (physical and biological) that are very generalized for their better practical applications:

PHYSICAL AND BIOLOGICAL CLASSIFICATIONS OF WATER

A general soil water classification system has been developed for a simple understanding of the soil water movement with respect to plant uptake. Plants are generally not capable of consuming water efficiently at both ends of the scale: 1) when the water in the soil is quite low with most of the pore spaces filled up with air, or, 2) when the water in the soil is flowing freely and the pore spaces are completely filled up with water.

To describe these phenomena on a biological scale, the terms unavailable, available, and superfluous are used. At the same time a measuring scale is used, which is based on tension or suction of water held in soil as soil water potential (expressed in bars or as Mega Pascal, MPa, in SI units. One unit of MPa is equal to 10 bars. These measurements give the physical classification of soil, namely gravitational water, capillary water, and hygroscopic water. When both the scales (Biological and Physical) are kept side by side showing the intensity of pull by soil for water under different energy levels (in bars, and are always expressed in negative values), one may have a general understanding of soil water measurement with respect to plant's uptake as shown below.

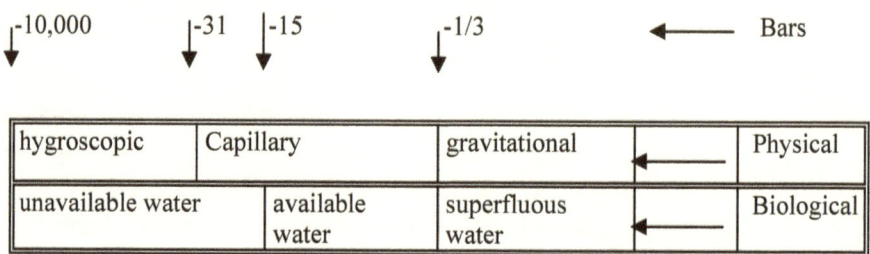

Soil moistures indicating -1/3 to -15 potential in the soil system are the indications for the available water for plant growth. This classification does indicate that with higher amount of soil moisture, the moisture tension shows lower negative value for the water potential.

SOIL TEXTURE IN RELATION TO SOIL MOISTURE

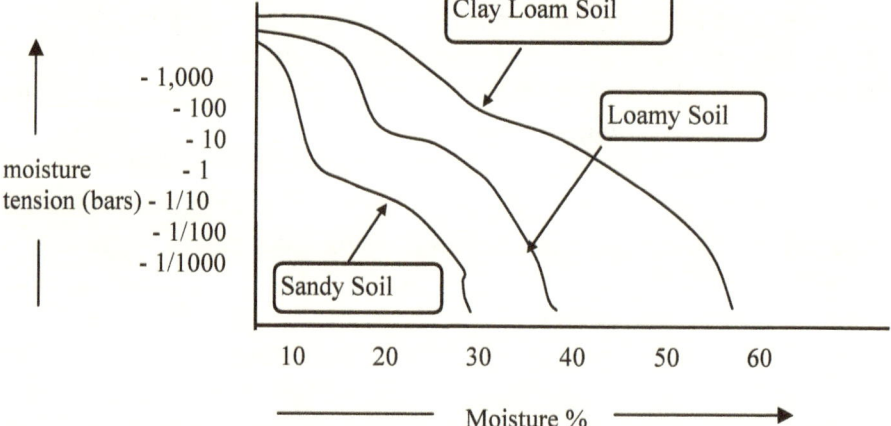

Fig. 3-2 : Soil textures in relation to moisture tension and moisture contents
Figure 3-2 shows that at the same moisture tension heavier soil will have more moisture content.

SOIL WATER MOVEMENT

(1) Unsaturated flow
(2) Saturated flow
(3) Water vapor

The capillary movement of water can best show unsaturated flow of water, viz.

h= 2T/rdg, where,

h = height in cm
T = surface tension in dynes/cm
r = radius of the capillary in cm
d = density in g/cc
g = gravitational acceleration cm/sce^2

However, it also includes the movement of water in soil due to difference in tension, which is useful for agricultural production. As such, the water always moves form low tension to high tension and feeds the plants. With gradual absorption of water at the rhizosphere, the moisture tension increases and the water flows towards the roots from the adjacent soil. This same system applies when the water moves in the soil system as water vapor. This makes the water vapor move from moist area to dry area as well as from warm area of soil towards the cold area of adjacent soil. This water vapor movement in soil system has immense importance, as most of the water movement in soil takes place under this phase, which does play an important role in plant growth.

METHODS OF SOIL MOISTURE DETERMINATIONS

1. GRAVIMETRIC METHOD:

Collect the soil samples from the field in plastic bags. Bring them in the laboratory and weigh. Then dry the soil at 105^0C and cool it. Weigh it again and calculate the % of moisture in soil on dry weight basis. A typical example is shown below:

Weight of moist field soil = 150 gms

Weight of dry soil = 139 gms
Weight of moisture = 150 - 139 = 11 gms
% Moisture in Soil = (11 x 100) / 139 = (1100) / 139 = 7.9136%

2. TENSIOMETER:

This measures the moisture tension but not the actual content of moisture in the soil. However, taking several readings of moisture tensions at varying moisture levels one can draw a standard curve for that particular soil and can determine the approximate moisture content of the soil at any time. This tensiometer is a useful tool for light to medium soil and as an aid to determining the irrigation time to maintain the moisture level within -1/3 to -15 moisture tension for plant's availability.

3. RESISTANCE BLOCK:

This is based on the principle that the electrical conductivity in soil system decreases with the decrease in soil moisture. Hence the conductivity in soil system decreases with the decrease in soil moisture. Hence the conductivity reading will be higher at higher moisture content as compared to less moisture content in the soil. This is more sensitive between -1 to -15 moisture tensions.

Actual moisture content (by gravimetric method) can be plotted against the conductivity readings to have a standard curve for any particular soil. If such standard curve is available for any soil, by the help of the conductivity readings, we can estimate the moisture content of that soil.

4. NEUTRON PROBE METHOD:

This probe contains the radioactive material containing 'the neutrons', which are made to move fast. However, in soil when they strike against hydrogen ions, they are slowed down and can be accounted by the counter. More slowed down neutrons when they are accounted by the counters indicate more moisture contents in the soil.

These counts, for any particular soil, can also be counted against the actual moisture content of the soil (by gravimetric method) and a standard curve can be made to know the moisture content if the counts of the hydrogen ions are known by this probe.

PROBLEMS AND SOLUTIONS TO UNDERSTAND SOIL MOISTURE USES

PROBLEM #1

A soil sample taken from a field was placed in an enclosed container and weighed, dried in oven at 105^0C and reweighed. Calculate the mass of water contained in the soil in percent if there were following measured values:

Moist soil plus the container = 160g, Oven dry soil at 105^0C + the container = 135g, Empty container weight = 40g

Solution: Moist soil = 160 - 40 = 120g

Oven Dried soil = 135 - 40 = 95g

Wt. of moisture = moist soil - oven dried soil = 120 - 95 = 25gms

% of moisture = 25 / 95 x 100 = 26.3157%

PROBLEM #2

If the bulk density of the soil in the above problem = 1400 kg/m^3, what will be the moisture percent by volume in that soil?

Solution: The volume of the water contained in that soil = Qv

Qv = (wt. of water) x (bulk density of soil)

(wt. of dry soil)   (density of water)

= (25/95) x (1400kg/m3/1000kg/m3)

= 0.26315 x 0.14 = 0.36841

Volume of water % = 0.36841 x 100 = 36.841%

55

PROBLEM #3

Given the following information, calculate
 (a) The total water present in top 30 cm
 (b) The depth to which 27.5 mm of irrigation water would wet this soil uniformly
 (c) The available water the soil contains in the top 30 cm, when the soil is at its field capacity.

The given information on the soil is: Percent water content = 15%, Water content at field capacity = 22%, Permanent wilting % = 8%, Bulk density of soil = 1300kg/m3 or (1.3g/cc)

SOLUTION:

 (a) PW (present mass of water content) = PW/100 = 15/100 = 0.15
Using the formula dw = Qv(ds) = Qm (Pb/PW)ds
Where
dw = depth of water
Qv = The volume of water content having the soil volume occupied by water,
      given as a fraction.
ds = soil depth.
Pw = Qm = water in mass which is contained
dw = Qm (bulk density of soil/density of water)ds
    = 0.15cm (1300kg/m3/1000kg/m3) 30cm
    = 1.3 x .15 x 30
    = 39 x .15
    = 5.85cm

(b) dw= 27.5mm

Qm = difference between present moisture and field capacity
      = (0.22 - 0.18 = 0.04)

This is because when water is added, it wets the soil from its present condition of 15% to its field capacity before all additional water (gravitational water) drains deeper.

Hence 27.5mm = (0.04cm water)   x   (bulk density of soil) ds
           (1cm depth of soil)   (density of water)

         = (0.04 cm) x (1300kg/m3) (ds in mm)
             (1)   (1000kg/m3)

or ds = $\dfrac{27.5}{0.04 \times 1.3}$ = $\dfrac{27.5}{0.052}$

or ds = 528.8 mm = 52.88 cm deep

(c) To calculate the total possible plant available water in the top 30cm when the soil is wetted, it equals field capacity minus the permanent wilting percentage, which is 0.22 - 0.08 in this example. Hence the plant available water is

= Qm (PB) ds

= (0.23 - 0.08) [(1300kg/m3) / (1000kg/m3)] x 30cm of soil

= 0.14 x 1.3 x 30 cm of available water in the top 30cm of soil

= 0.182 x 30 = 5.46cm of available water

CATION EXCHANGE CAPACITY (CEC) OF SOIL

The cations in solution when they are exchanged with other cations held at the surface of any surface active material like organic matter or the finer soil separates like clay or silt in the soil system it is known as cation exchange. Thus adsorption capacity of any known amount of soil can be determined with reference to the exchangeable cations available in that soil system. The active soil material can be said to have higher, medium or lower capacity. Such chemical measurements can be expressed in centimoles per kilogram (c mol/kg) or in meq/100g of soils. Such measurements and evaluation for any soil may indirectly reveal its inherent productive capacity and may be a very valuable tool in soil fertility measurements. Because naturally those soils, which can adsorb more cations, are supposed to have the capacity to exchange more and may possess better productive capacity if managed properly.

A background information may be necessary to fully comprehend this vital phenomenon of CEC in soil:

(a) Definitions of mole, centimole, and CEC

(b) Determining the charge and subsequent calculation

(c) CEC under natural condition

(d) CEC in relation to soil pH

(e) Soil colloids, including silt and clay with their existing charges

(A) MOLE AND CENTIMOLE:

A mole weight is simply called a mole of a substance having 6.023×10^{23} items including molecules, ions or atoms. However, one atomic weight or one formulae weight of a substance will have 6.023×10^{23} molecules, ions or atoms. Hence the mole weight of K, Ca, and H will be 39, 40 and 1. However for a compound like KCl, the mole weight will be $39 + 35 = 74$ gms which will possess 6.023×10^{23} molecules of KCl. But we are interested in the number of positive charges that a colloid like particle of soil having negative charges can hold. But the moles held by 1 kg of soil will have small numbers like 0.24 moles/kg of soil. Hence centimoles (1 mole = 100 centimoles) have been adopted which will give 24 centimoles/kg of soil or 24 c moles/kg.

For example if a soil has CEC of 10 c mole/kg of soil means that 1 kg of this soil can adsorb 10 c moles of H+ ion and can exchange for 10 c moles of monovalent ions like K+ and Na+ or 5 c moles of divalent ions like Ca++ and Mg++ alone or in combination. Because, these exchanges do take place as based on the chemical equivalence.

(B) DETERMINING THE CHARGE, THE SUBSEQUENT EXPRESSION AND CALCULATIONS

The following steps may be necessary for this:
(a) Measure the amount of each common ion in one kg of soil
(b) Calculate the number of centimoles of charge it gives Table 3-2.

Table 3-2: Shows the Charges and Weights of One Mole of Ion or Compound.

Ion or Compounds	Wt. of one mole	Wt. needed to give one mole of charges	Wt. for one centimole of charge
$CaCo_3$	100	100/2 = 50	0.50 g
Ca^{++}	40	40/2 = 20	0.20 g
Na^+	23	23/1 = 23	0.23 g
NH_4^+	18	18/1 = 18	0.18 g
H^+	1	1/1 = 1	0.01 g

Now if any soil contains 0.6 g of exchangeable Ca^{+2} per kg of soil. There is then from the above table we find 0.2 g of Ca^{+2} gives a charge of 1 c mole. Then this soil occupies 3 c mole of negative charge sites. Similarly other cations can also be calculated and their sum will give the CEC of soil.

Thus, when you keep the chemical equivalence in your mind, you can express cation exchange in field-terms. The good example of this will be when you lime an acid soil and when Ca^{+2} ions replace the part of H^+ ions as indicated below, where micelle represents a clay particle.

Thus two H^+ ions are replaced by the equivalent charge associated with Ca^{+2} ion. Thus 1 mole of H^+ (1 g) would exchange 1/2 mole of Ca^{+2} ion (40/2 = 20 g). Similarly, to replace 1 centimole H^+/kg would require 20/100 = 0.2 g Ca^{+2}/kg soil. On the same basis, the amount of Ca^{+2} needed per hectare furrow slice (2.2 million kg of soil) is 0.2 x 2.2 x 106 g or 440 kg. This will amount to 440 x 2.5 = 1100 kg of limestone as the ratio of limestone/Ca = 2.5. And thus, 1100 kg of limestone/hectare of soil will exchange 1 mole of H^+/kg of soil.

(C) CATION EXCHANGE UNDER NATURAL CONDITION

Under natural conditions H^+ generated by the organic matter decomposition or by the breakdown of carbonic acids may always tend to replace other adsorbed cations from the colloidal complexes of soil (micelle) which make the reaction to move towards right. However when Ca^{+2} is applied in the system the reaction may move to the left. In each case, the equivalent amount of Ca^{+2} is replaced by H^+ or vice versa.

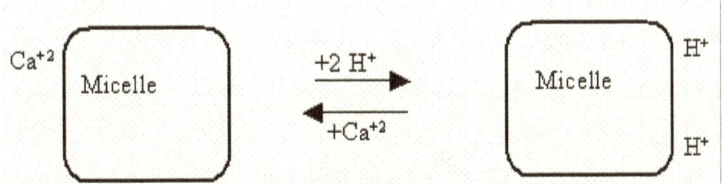

However, the above example is very simple and does not fit the complexity of soil, where simultaneously other cations are also involved and H+ in the system are being generated by the carbonic acid (H_2CO_3) to push the reaction to the right with higher precipitation but may move to the left under xeric condition.

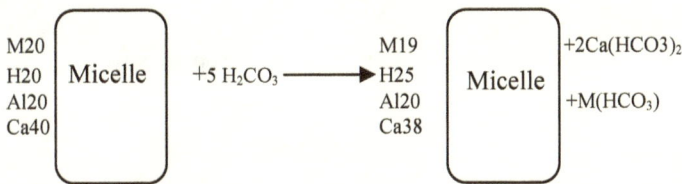

These reactions always proceed with or without fertilizer applications based on soil's capacity to exchange cations under prevailing conditions.

To determine the value of CEC in c moles/kg of soil the following background knowledge may be necessary in continuation of previous Table 3-2.

Ion or Compound	Wt. of one mole	Wt. needed to give one mole of charge	Wt. for one c mole of charge (g)
H^+	1 g	1 g	0.01 g
NH_4^+	18 g	18 g	0.18 g
Na^+	23 g	23 g	0.23 g
Ca^{++}	40 g	40/2 = 20 g	0.20 g
$CaCo_3$	100 g	100/2 = 50 g	0.50 g

PROBLEM #1

If 1 kg of soil contains, 8 c mole of Ca^{++}, 4 c mole of Mg^{++}, 1 c mole of K^+, 1 c mole of Na^+, 1 c mole of NH_4^+, 1 c mole of Al^{+++} and 4 c mole of H^+ as the entire exchange, what is the CEC of this soil?

SOLUTION:

CEC = 8+4+1+1+1+1+4 = 20 c mole/kg of soil

PROBLEM #2

If all the charges of one kg of soil have been neutralized by the following adsorbed cations: Ca^{++} = 0.8 g, Na^+ = 1.0 g, H^+ = 0.04 g.
What is the CEC of this soil?

SOLUTION:

If 0.2 g of Ca^{++} gives 1 c mole of charges,
  then 0.8 g of Ca^{++} will be = 4 c mole of charges.
  Similarly 0.23 g of Na^+ gives 1 c mole of charges,
  then 1 g of Na will give 1/0.23 c mole of charges = 4.347 cmole of charges.
  Similarly .01 g of H^+ gives 1 c mole of charges.

Hence 0.04 g of Hydrogen will give = 1/.01 x 0.04 g mole of charges = 0.04/0.01 = 4
cmole of charges.

Hence the total charges in one kg of soil will be (4 + 4.347 + 4) cmoles/100 = 12.347 c moles/kg of soil. This is the CEC of the soil = 12.347 c mole/kg.

PROBLEM #3

Charges of basic cations are 30% and total CEC is 45. What is the c mole of charges for this basic cation.

% base saturation = (c mol of charges for basic cations) x 100

Total CEC

30/100 = c mol of charges for basic cations
            45

hence c mole of charges for basic cations = $\frac{30 \times 45}{100}$ = $\frac{1350}{100}$

= 13.50 c moles/kg

PROBLEM #4

What is the mole charge weight and c mole charge weight of each of the following:

SOLUTION:

Ions	Mole Wt.	Wt. needed for 1 mole of charge (+)	Wt. needed for 1 c mole of charge (+)
Ca^{++}	40	40/2 = 20 g	0.20
Mg^{++}	24	24/2 = 12 g	0.12
K^+	39	39/1 = 39 g	0.39

PROBLEM #5

What is the weight of 1 c mole of (positively charged) Ca^{++}/kg of soil in 1000 g of soil, in ppm, as well as in lbs/acre?

1 c mole of Ca^{++} = 40/2 g = 20 g

1 mole of charge = 20/100 = 0.2 g/kg of soil

Which will be 0.2 x 1000 g/1000 kg of soil

= 200 g/1000 kg of soil

Which will be 200/1000 kg/1000 kg of soil

= 0.2 kg/1000 kg of soil

0.2 kg x 1000/1000 x 1000 kg of soil

= 200 kg/1,000,000 kg = 200 ppm = 200 lbs/1,000,000 lbs of soil

= 400 lbs/2,000,000 lbs of soil

= 400 lbs/acre of soil

(One acre furrow slice of soil, 7 inch deep, weighs about 2,000,000 lbs by weight. Hence while converting ppm values, we multiply by 2 to get lbs / acre)

It also appears that soil texture; clay content, type of clay organic colloids and weathering intensity are the controlling factors to influence low or high CEC of a soil. In this respect, soil texture and clay content have direct relationship, because heavier textured soil will always have more clay content, which reflects an increased CEC. Expanding type of 2:1 type clay has higher CEC than non-expanding 2:1 type clay and fixed 1:1 type clay. And the least CEC in soil is influenced by the intense weathering as well as by clay composed of oxides of Fe and Al. The following table reflects the above discussion as has been found in the soil system.

Table 3-3 : The relationship of various soil suborders, clay type, texture and CEC at different locations.

Location	Soil Texture	Soil Suborders Classified	CEC moles/kg	Possible Readings
NJ	Sandy	Udult (more weathered)	2	Light Texture,
WI	Sandy	Psamment (young)	3	Light Texture
NJ	Sandy Loam	Udult (more weathered)	3	Light Texture
SC	Sandy Loam	Udult (more weathered)	6	Medium Texture
NJ	Loam	Ochrept (young soil)	11	Medium Texture
WI	Silt Loam	Udalf	23	Heavier Texture Less weathered Clay (2:1 dominant)
AL	Clay	Udult	4	Highly Weathered Clay (1:1 and oxides)

(D) CEC IN RELATION TO SOIL pH

In general, CEC in soil has been observed to be pH dependent. At higher pH above 6.5 the CEC in the mineral soil suddenly shows an increase as the pH increases. While below pH 5.5, CEC is lower and does not reflect any change by lowering the pH. However in case of organic soil the cation exchange capacity does show a gradual increasing trend with increasing soil pH irrespective of the pH level.

(E) SOIL COLLOIDS INCLUDING SILT AND CLAY WITH THEIR EXISTING CHARGES

Any particle in soil system, which is less than 1 micron (0.001 mm), is known as colloid. Hence, same portions of clay and organic matters or molecules may be less than 0.001mm and are known as colloids. These particles including the finer particles like silt (.002 - .005 mm) or clay (.002 mm or less) may possess negative charges due to the following reasons:

1. The dissociation of OH $^-$ ion in the soil system may generate the extra negative charge

$$OH^- \longrightarrow O^= + H^+$$

2. The dissociate of OH $^-$ from an organic colloid

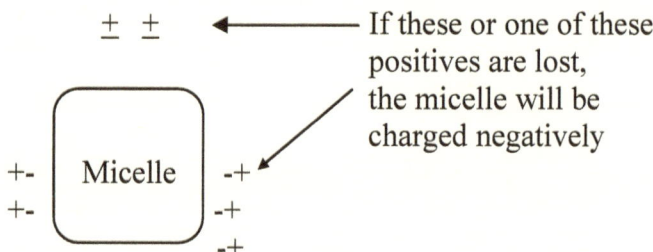

3. Due to the loss of cations from the dynamic soil system as shown:

$$\underline{+}\ \underline{+} \longleftarrow \text{If these or one of these positives are lost, the micelle will be charged negatively}$$

+- | Micelle | -+
+- | | -+
 | | -+

This entire system is just like a double layer and supposing some of the positive charged ions are lost from these swarms of cations, the negative charge will be exposed and the clay particle will gain negative charge.

4. By Isomorphous substitution as shown

(a) If, Al^{+3} with three charges, is substituted by the Mg^{+2}; there will be gain of negative charge

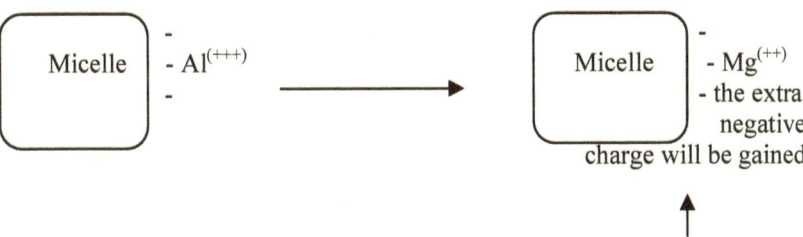

(b) Similarly if Si $^{++++}$ is substituted or replaced by Al $^{+++}$. The gain of extra negative charge will take place.

These phenomena are the most important which create the possibilities of generating cation exchange capacities in the soil system.

NUTRIENTS FOR PLANT GROWTH

There are sixteen essential elements, which are needed for the normal growth of plants. They are essential, because in lack of any one of these, the plants will not achieve their normal growth. Their main sources are either water, air, or soil. Nine of these essential elements are needed in larger quantities and as such, these are known as major or macronutrients. Other seven elements are needed in very minute quantities and, therefore, are known as micronutrients or minor elements for plant growth. However all are essential. Besides, there are certain other elements, which are known as beneficial elements. These beneficial elements may be essential for only certain plant-species or may stimulate growth in plants, but are not essential for all plants. The following tables will clearly illustrate the above statements:

Table 3-4: Essential Elements, their Chemical Symbols and their Available Forms.

(I) Macronutrients from air and water	Symbols	Available Forms
(1) Carbon	C	CO_2, HCO^{-3}
(2) Hydrogen	H	H_2O, NH_4^+, $H_2PO_4^-$, $HPO_4^=$
(3) Oxygen	O	HCO_3^-, $H_2PO_4^-$, $HPO_4^=$
(II)Macronutrients from soil		
II (a) Primary fertilizer elements		
(4) Nitrogen	N	NH_4^+, NO_3^-
(5) Phosphorus	P	$H_2PO_4^-$, $HPO_4^=$
(6) Potassium	K	K^+
II (b) Secondary (amending elements)		
(7) Calcium	Ca	Ca^{++}
(8) Magnesium	Mg	Mg^{++}
(9) Sulfur	S	$SO_4^=$
II (c) Others		
(10) Copper	Cu	Cu^{++}
(11) Manganese	Mn	Mn^{++}
(12) Iron	Fe	Fe^{++}
(13) Zinc	Zn	Zn^{++}
(14) Boron	B	$B_4O_7^=$
(15) Molybdenum	Mo	$MoO_4^=$
(16) Chlorine	Cl	Cl^-

Table 3-5: Beneficial elements, their chemical symbols, available forms, as needed by special plants.

Name	Symbol	Available form	Needed By Special Plants
Sodium	Na	Na^+	Halophytes, celery, spinach
Cobalt	Co	Co^{++}	Legume bacteria
Silicon	Si	$H_3SiO_4^-$	Rice, sugarcane
Nickel	Ni	Ni^{++}	Legume (activates urease activity)
Aluminum	Al	Al^{+++}	tea plants (stimulates growth)

It is interesting to note that most of the constituents of living things are basically made of H, C, N, O, P and S. Besides, it has also been observed that H, C, N and O are most abundantly available elements in the cosmos. And H has a unique chemical characteristic to combine with the above elements forming a basic structural material for all living things including plants of this universe. The following table briefly signifies their importance in living plants:

Table 3-6: The role of H, O, C, N, and P to form their level of organizations

# Of Elements	Carrier of				
	Organization	Matrix	Energy	Regulation	Information
	H	O	C	N	P
1	H				
2	H_2	O			
3	H_x	O_y	C_z		
4	H_x	O_y	C_z	Nq	
5	H_x	O_y	C_z	Nq	Pr

With the above background, we can list the main functions of all the sixteen essential elements in plant-growth and may note down the major symptoms when they are deficit and the right soil pH, which helps plant in uptake of specific nutrients.

Table 3-7: Essential elements relating their main functions in plants, deficiency symptoms, and soil pH requirements for efficient plant uptake.

Essential Elements	Their Main Functions	Signs of their deficiency in plants	Soil pH Range needed for availability
C	Photosynthesis & Energy	No growth	3-11
H	Organization	No growth	3-11
O	Aeration & Matrix	No growth	3-11
N	Regulation, growth of protein, and enhancement of green coloration	Yellowing of leaves from older leaves starting by the mid vein from the tip towards the stem	5.5-8.5
P	Genetical information, & fruiting, and root formations	Reddish purple leaves in younger plants on the edges of the leaves.	6-8
K	Makes it resistant to diseases & lodging of plants. Aids photosynthesis and translocations of photosynthates	Browning of the edges of the lower leaves and root reduction	6-10
Ca	Promotes early root growth and corrects soil's acidity	Poor root growth and rotting after blackening of the main root	6-10
Mg	Maintains chlorophyll & dark green color and regulates the uptake of plant nutrients & corrects soil's acidity	Shows up on lower leaves with main veins remaining green but within the veins the bronze color appears.	6-10
S	Increases root growth and nodulation in legumes. Makes plant vigorous & corrects soil alkalinity	Stunted plant growth chlorosis on leaves with redness on lower leaves as well as on younger leaves.	3-8

Table 3-7 (Continued)

Essential Elements	Their Main Functions	Signs of their deficiency in plants	Soil pH Range needed for availability
Fe	Increases chlorophyll and enzyme activities	Chlorosis of younger leaves with contrasting green veins and yellow tissues.	3-7
B	Healthy growth of pollens and pollen tubes aids in cell wall and seed formations.	Growing points and younger leaves are stunted.	4.5-7.5
Mn	Activates chlorophyll and carbohydrate formation	Yellowing of younger leaves with brownish black specks.	4-7
Zn	Activates chlorophyll, and growth of hormones.	Younger leaves become pale And white bud on the growing points may form.	4.5-7.5
Cu	Acts as a catalyst and regulator in chlorophyll formation.	Dieback, chlorosis, and loss of leaves' turgidity.	4.5-7.5
Mo	Helps rhizobium in legumes & synthesizes proteins.	General stunting of plants.	6-10
Cl	Helps in enhancing the quality of tobacco leaves.	Not clear	Not known

Based on the table as well as on the crop needs, it is easily visualized that 6.5 pH in general is suitable for most of the crops. However, there are some crops (especially non-legumes) which grow better in quite an acid medium (5.5 to 6.5 pH) while others (which are legumes) grow better in neutral to alkaline medium (7 to 8.5 pH). This is also imperative and well understood that legumes generally like to have divalent cations (Ca^{++} and Mg^{++}) while the plants of graminaceae family like monovalent cations (K^+) for their growth.

The deficiency sign of any element when it becomes obvious in the plants, the yield if any you have already lost will be hard to correct during that short growing period of that crop. However, for the next year and in fruit trees for later years suitable measures can be taken to correct the deficiency to enhance better yields. On a regular basis, soil testing is essential to apply fertilizers judiciously because excess applications are environmentally as well as economically more harmful than fewer applications.

BIBLIOGRAPHY

1) Brady, Nyle C. 1990. The Nature and Properties of Soils Tenth Edition. Macmillan Publishing Company New York.

2) Dunahue, Roy L., Follett, Roy Hunter, and Tulloch, Rodney. 1983. Our Soils and Their Management Sixth Edition. Interstate Publishers Inc. Danville, Illinois.

3) Marschner, Horst. 1986. Mineral Nutrition of Higher Plants. Academy Press. USA.

4) Miller, Raymond W., and Donahue, Roy L. 1990. Soils An Introduction to Soils and Plant Growth. Sixth Edition Prentice Hall Inc. USA.

5) Pepper, I.L. 1996. A biotic Characteristics of Soil. Pollution Science. p. 9-18. Academic Press Inc. Ed. By Pepper, Ian L., Gerba, Charles P., and Brusseau, Mark L.

6) Singer, Micheal J., and Munns, Donald N. 1987. Soils An Introduction. Macmillan Publishing Company. USA.

7) Sprague, H.B. 1964. Hunger Sign in Crops. 3rd edition. p. 461. New York David McCoy Co. Inc. USA.

8) Soil Constraints on Sustainable Plant Production In The Tropics. 1991. Tropical Agricultural Research Series No. - 24. Proceedings of the 24th International Symposium, Kyoto, Japan.

9) Waring, Richard H., and Schlesinger, William H. 1985. Nutrient Uptakes and Internal Distribution: The Intra-system Cycle Forest Ecosystem Concept and Management. p. 157-180.

10) Wild, A., Jones, L.H.P., and Macduff, J.H. 1987. Uptake of Mineral Nutrients and Crop Growth: The Use of Flowing Solution. Advances in Agronomy 41:171-215.

CHAPTER 4

Mineralogical Properties of Soil

Minerals may be soft or hard which have been defined as a solid mass possessing a characteristic chemical composition consisting of limited range of compositions in a systematic three dimensional atomic order. These are either, homogeneous in their chemistry and physical properties or exhibit with restricted systematic variations irrespective of being naturally and inorganically produced. However, those minerals having regular, ordered atomic structures are said to be crystalline. On the other hand, those naturally occurring solid or liquid that lacks systematic arrangement of constituent atoms are known as mineraloids and are non-crystalline or amorphous. For example, volcanic glass, coal and petroleum are mineraloids.

Basically, all these minerals come out of rocks, which are naturally occurring, coherent, multigranular aggregate of one or more minerals and/or mineraloids. These rocks (mixture of minerals and/or mineraloids) are basically grouped into three major kinds: (1) Igneous, (2) Metamorphic and (3) Sedimentary. Igneous rocks are those where molten silicate minerals or magma solidify into a combination of glasses and minerals. These rocks constitute 80% of the earth's crust and the examples of these rocks may be granite, diorite, gabbro, rhyolite, andesite and basalt. Metamorphic rocks are those rocks whose original minerals/or textures have been markedly altered by recrystallizations or deformations. Soapstone, marble, and slate are the representative examples of metamorphic rocks. Sedimentary rocks have resulted from the deposition and recementation of weathering products of other rocks. Examples of sedimentary rocks are: limestone, dolomite, sandstone, shale and conglomerate.

Soils are formed from the parent material, which may be either formed in situ or formed from the transported material or organic residues. The weathering of rock to smaller particles forms the parent materials. Jenny (1941) was the first American soil scientist who hypothesized that soil, a dependent variable, was a function of five independent soil forming factors:

$$S = f(P, C, R, B, T)$$

Where S (soil) is a function of the parent material (P), climate (C), topography or relief (R), biota (B), and time (T).

The parent material determines the soil mineral composition. Climate determines the temperature, and amount and rate of precipitation whereas topography determines the amount of water entering the profile. A normal soil contains some organic matter and the amount and type of organic matter depends on biota. By time we mean here the time required to develop horizons. Under favorable conditions, it may take 200 years, but under less favorable conditions it may take thousands of years. The soil forming factors are inputs of soil formation, which go through different soil forming processes. During the process of soil formation bigger rock particles are broken down to small pieces through physical and chemical weathering such as freezing and thawing, uneven heating, shrinking and swelling. Once the rocks are broken, their mineralogy is changed through hydrolysis, hydration, oxidation-reduction, dissolution and/or combination of these. Through water, the weathered and chemically changed/modified material is moved either downward or upward. Through downward movement or leaching the ions, clay, and mineral particles form subsoil. Upward translocation occurs in semi-arid area where evapotranspiration exceeds precipitation and salt moves on surface layer during summer time.

MINERAL FORMATIONS

PRIMARY AND SECONDARY MINERALS

Minerals may be either primary or secondary. Primary minerals are those that are formed from molten lava and have not been altered chemically since deposition and crystallization such as feldspars (orthoclase and plagioclase), micas, or apatite. On the other hand secondary minerals result from decomposition of primary minerals by passing through phase change. Phase change simply means that the original solid materials have to pass through liquid or gaseous stage. This can be accomplished by reprecipitation of decomposed gaseous or liquid products of primary minerals, forming clay or clay-like materials. Some minerals such as quartz, calcite, dolomite, and gypsum may be primary or secondary minerals. During the process of soil formation, primary minerals may generally change into secondary clay minerals. Thus they change from solid to liquid or gas and resolidify or reprecipitate.

MAJOR GROUPS OF SOIL MINERALS

Soil minerals; mainly consist of silicate minerals that are crystalline or amorphous. In crystalline silicate minerals, silicon (Si) and or aluminum (Al) and oxygen (O) atoms are arranged in a definite pattern but not in amorphous minerals. In earth's crust Si, O, and Al are found in the proportions of 47, 27 and 8 percent, respectively. One atom of silicon combines with 4 atoms of O and forms a tetrahedron whereas one atom of Al combines with 8 atoms of oxygen and forms an octahedron. In soils many tetrahedral and octahedral form tetrahedral and octahedral sheets, respectively and these sheets are joined together to form one unit layer which is stacked on top of one another to form different clay minerals. The space between units is called interlayer, which are major locations of cation exchange. Based on the number of sheets in the unit, layer silicate minerals are divided into 4 major groups: (a) 1:1 fixed, (b) 2:1 expanding, (c) 2:1 and 2:1:1 non-expanding (d) oxides of Fe or Al or combination of both. Kandites are 1:1 fixed type (where silicon and aluminum sheets are in ratio of 1:1) with no expansion capability. Smectites and vermiculites are expanding types with 2:1 ratio of silicon and aluminum sheets. A 2:1 non-expanding type mineral is illite; which is due to potassium restricted in expansion. Chlorite has 2:1:1 ratio of silicon, aluminum sheets but is not an expanding type due to the presence of brucite $Mg(OH)_2$ layer in between. The above minerals are generally crystalline. But, the oxides of Fe or Al are amorphous minerals with negligible CEC values. Clay minerals have net negative charge, which comes from ionization of hydrogen ions forming hydroxyl ions on clay surfaces and also from isomorphous substitution. Isomorphous substitution is substitution of one ion for another of similar size but lower valance. For example, substitution of Al^{+3} for Si^{+4}, or Fe^{+2} and Mg^{+2} for Al^{+3}, gives rise to net negative charge in the clays that will attract and hold more positively charged ions (cations).

Table 4-1: Characteristics of common clay minerals showing their expandability, CEC, and the environment in which they are formed.

Mineral	Expandability	CEC (meq/100 g soil)	Environment for its formation
Kandites	None	3-15	Moist, warm hot, sub-humid and humid leached soils
Smectites	High	60-100	Arid to humid soils having limited leaching
Vermiculite	High	80-150	Sub-humid to humid soils high in micas
Illite	Low	20-40	Sub-humid and cool areas soils high in micas
Chlorite	None	2-5	Clays previously formed in mica's sediments
Sesquioxides (ex. Fe2O3 + Al2O3)	None	0-3	Wet, hot, highly weathered tropical soils
Amorphous (ex. Allophane)	None	50-150	In rapidly weathering, young volcanic ash

Kandites: Kandite is a group of clay mineral having one sheet of silica tetrahedral per corresponding sheet of alumina octahedral (1:1), which includes minerals such as nacrite, dickite, halloysite, as well as kaolinite. There is very little substitution of Al^{+3} for Si^{+4} or Mg^{+2} for Al^{+3}, so the net negative charge is low. In kaolinite, the hydroxyls from the surface of one unit cell face the oxygen atoms from the opposite face of next unit cell and form a weak hydrogen bond. These weak hydrogen bonds cumulatively form strong bonds and keep the sheets together tightly so that water and ions cannot penetrate into the interlayer.

Smectites: They are 2:1 type minerals having 2 sheets of silica tetrahedral per alumina sheet per layer. The most important mineral in this group is montmorillonite; other members include nontronite, saponite, hectorite, and biedellite. Unlike kandites, in smectites, oxygen of silica tetrahedral face each other across interlayers without any hydrogen bonding which allows the movement of water between interlayers. Due to isomorphous substitution of Al^{+3} for Si^{+4} and Fe^{+2}, Mg^{+2} or Ca^{+2} for Al^{+3} there is generated a net negative charge in smectite. Thus a higher cation-holding capacity of smectite results into an increase swarm of cations with very little loss of cations due to leaching. Presence of water molecules and cations in the interlayer makes smectite expanding type mineral and the unit cell size may vary form 0.96 to 1.8 micron or more.

Illite: It is also called hydrous mica because it is derived from mica. It is 2:1 type of clay mineral similar to smectite with K as interlayer cation. However, K^{+} ion holds layer so tightly that water is restricted to enter freely into the interlayer, which makes it nonexpanding.

Vermiculite: Vermiculite is a 2:1 type of mineral in which substitution occurs mostly in tetrahedral sheet. The interlayer is occupied by hydrated magnesium, which holds vermiculite layers more tightly than smectites and results in less expansion than smectite.

Chlorites: Chlorites are 2:1:1 type of clay minerals i.e. each layer of chlorites has 2 sheets of silica tetrahedral, 1 sheet of alumina octahedral, and 1 sheet of magnesium octahedral {(Mg) 6 (OH) 12, brucite}. Al^{+3}, Fe^{3+}, and Fe^{+2} can be substituted for Mg^{+2} in the brucite sheet, which can result in a net positive charge and sheets can be held together firmly.

Sesquioxides or non-silicate clays (metal oxides and hydroxides): In hot tropical climates under high rainfall and long-time intensive weathering, most of the silica and alumina get dissolved and leached away. The remaining material on surface horizon is named as sesquioxide. They may be well crystalline minerals, noncrystalline or poorly crystalline. Crystalline oxides include oxides and hydroxides of iron such as hematite (Fe_2O_3) and goethite (HFeO2) and oxides and hydroxides of aluminum such as gibbsite {(Al(OH)3} and boehmite (AlOOH). Amorphous clays are mixture of Si and Al but have not well defined crystals and are not classified as minerals. They are sometimes referred as allophane because they look like a gel. Because of noncrystallinity they are not well characterized. The charges in oxides and hydroxides, whether crystalline or amorphous are generated form hydroxyl ions (OH)⁻ and can vary depending upon soil pH (variable or pH dependent). At high pH the H^+ from hydroxyl group dissociates that results in high cation exchange capacity but at low pH the OH group may get protonoted and can result in anion exchange capacity.

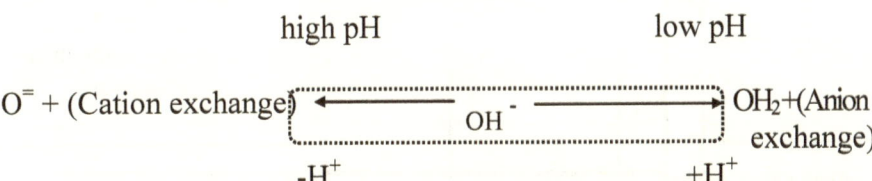

A small amount of these clays exists in many soils but are most dominant in humid, hot, and well drained soils of tropics. These clays do not swell or stick. Such changes can briefly be summarized by depicting the possible routes in the formation of silicates and oxide clays as weathering proceeds under mild and intense weathering conditions under any given time frame.

WEATHERING OF MINERALS

In nature, minerals are formed and weathered. This always continues under any given conditions. However, it is in the weathering mode that nutrients for growing plants get released from the minerals, as they move from primary minerals to secondary minerals:

Primary Minerals   **Secondary Minerals**

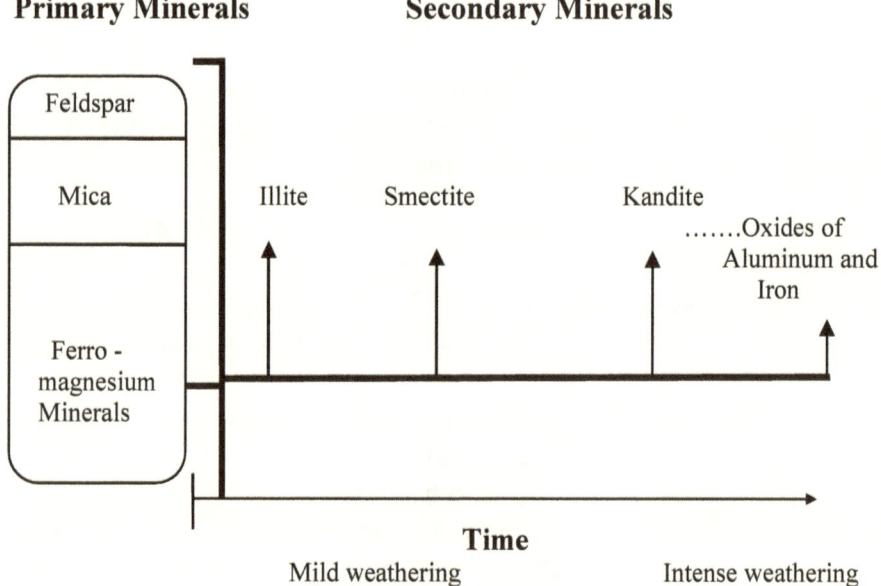

Fig. 4-1: General path of mineral weathering with time

The earlier discussion on the mineralogical properties of soil clearly identifies one important aspect of soil, which indicates that irrespective of chemical composition of soil, mineralogical aspects of soil may very well control the gradual, rapid or no release of plant nutrients. For example the presence of biotite in soil may assure the release of K^+, however there is no immediate assurance of the release of K^+ from the potassium feldspar like microcline. Hence, the respective resistance of minerals to weathering directly controls the release of nutrients in the soil system. Such resistance can be summarized as follows:

1. The minerals that are most stable at high temperatures are the least stable at low temperature if they contain alkaline earth cations (Ca or Mg). Examples are calcitic feldspar, olivine and hypersthene.

2. Increased links between the tetrahedron resists weathering. Examples being feldspars and quartz, where each of the four tetrahedral oxygen are in the corners of other tetrahedral.

3. Position of ions and loose fitting of ions make the mineral structure weak. For example, calcium feldspar is far less stable than potassium feldspar.

4. Charge imbalances may also create instability of minerals. When Mn^{+2} or Fe^{+2} change to Mn^{+3} and Fe^{+3} in aerobic conditions, they are prone to weather comparatively quicker. But in anaerobic conditions, oxidized ions of Fe^{+3} and Mn^{+3} are also comparatively quite unstable.

Tables 4-2 and 4-3 explain the above statement justifying the comparative resistance of minerals to weathering, which is based on their mineralogical structures in nature. However, in general, it is the change in soil moisture and temperature regime that makes the minerals prone to weathering. But even then, some minerals are structurally more stable than the others. The table, that follows, may illustrate this.

Table 4-2: Weathering index of primary minerals based on their structural orientations

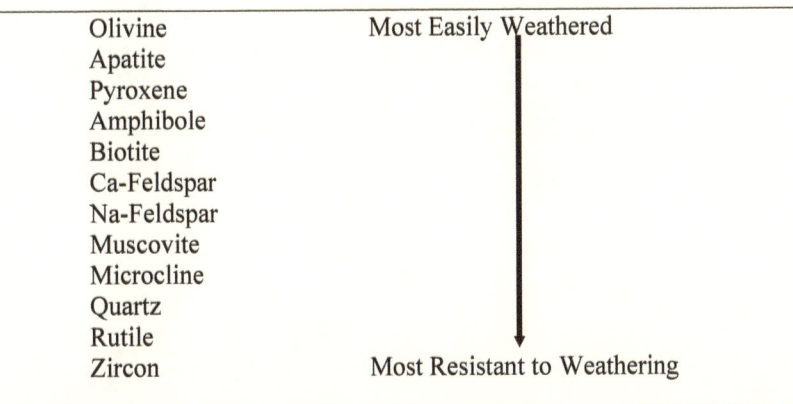

Olivine	Most Easily Weathered
Apatite	
Pyroxene	
Amphibole	
Biotite	
Ca-Feldspar	
Na-Feldspar	
Muscovite	
Microcline	
Quartz	
Rutile	
Zircon	Most Resistant to Weathering

Table 4-3. Weathering index of secondary minerals based on their structural orientations

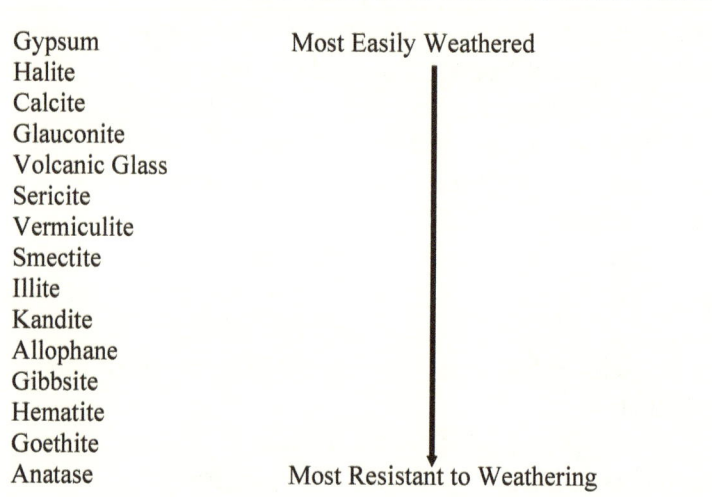

Gypsum	Most Easily Weathered
Halite	
Calcite	
Glauconite	
Volcanic Glass	
Sericite	
Vermiculite	
Smectite	
Illite	
Kandite	
Allophane	
Gibbsite	
Hematite	
Goethite	
Anatase	Most Resistant to Weathering

BIBLIOGRAPHY

1) Berry, L G., and Mason, Brian. 1959. Mineralogy Concepts Descriptions, Determinations. W.H. Freeman and Company. USA.

2) Bohn, McNeal and O'Connor. 1979. Stability of Minerals in Parent Material. p. 115-117. Soil Chemistry A Wiley-Interscience Publication. John Wiley and Sons. N.Y.

3) Brady, Nyle C. 1990. Soil Colloids; Their Nature and Practical Significance. p. 186-196. The Nature and Properties of Soils: Tenth Edition. Macmillan Publishing Co. N.Y.

4) Ernst, W. G. 1969. Earth Materials. Foundation of Earth Science Series. Prentice Hall, Inc. New Jersey.

5) Foth, H. D. and Ellis, B.G. 1976. Soil Fertility. John Wiley and Sons.

6) Grim, Ralph E. 1968. Clay Mineralogy: Second Edition. McGraw-Hill.

7) Jenny Hans,. 1941. Factor of Soil Formation. A Classic of the Soil's Literature. McGraw-Hill. New York.

8) Mineral Classification of Soil. 1985. Publication #16. Soil Science Society of America Inc. American Society of Agronomy, Inc. Madison, Wisconsin.

CHAPTER 5

Biological Properties of Soil

The physical, chemical, and mineralogical properties of any soil are constantly influenced by their biological properties within the existing ecosystem. The impact of biological phenomena is so great that the soil mass has always been apprehended as a living identity itself.

SOIL AS A HARBOR FOR MICROBES

Trillions of plant floras (Fungi, Actinomycetes, Algae, and Bacteria) thrive in the soil system, which have always been associated with the living as well as the dead plant materials. There also exist various animal fauna (squirrels, gophers, mites, insects, beetles, slugs and snails) as macro-herbivores and detrivores. In addition, various micro animal faunas (rotifers, nematodes, and protozoa) do exist to a great extent as detrivores, predators or parasites in soil. Having innumerable microbes intertwined with the non-separable macro as well as microscopic organisms, this system of soil as a whole is capable of exchanging as well as energizing the entire soil mass to keep it as a living entity.

In spite of their beneficial effects, there are some soil worms like nematodes, cutworms, and wireworms, which destroy the economic plants. However, the beneficial effects of others outweigh the ill effects caused by these. Most of the microbes and insects enhance the aggregation properties of soil. Some fix nitrogen from the atmosphere, some convert nitrogen within the soil system to make nitrate forms available for plants. Actinomycetes may fix nitrogen and are considered to be a major source of antibiotics used for controlling diseases. They also accelerate the process of 'humus' formation, which may be called the end product of organic matter decomposition. Basically, bacteria and fungi are the most important categories of microbes to activate organic matter decomposition, nitrogen fixation, as well as to react as oxidizing or reducing agents. In addition, these microbes may also cause soil and plant diseases.

RHIZOSPHERES

The area in the soil system, close to the roots with vital influence on root-mass, is known as rhizosphere. There, the active concentration of microbes always exceeds manifold as compared to neighboring soils. Their activities closer to roots makes the soil pH lower due to their metabolic activities resulting into the constant release of CO_2 in the soil system increasing mild acidity ($CO_2 + H_2O = H_2CO_3$)

In this rhizosphere, the active roots do secrete various combinations of organic acids, proteins, growth substances, growth inhibitors, repellents as well as attractants for the microbes. These combinations of activities make the rhizosphere a very complex system.

HUMUS

Humus is a lignoproteinous material, an end product of organic matter decomposition, which is dark, heterogeneous mass, and is always in a dynamic stage of very gradual change. Physically, 'Humus' is a glue like colloidal material which roughly contains 40 to 45% Lignin, 30 to 35% protein, and the rest 20 to 30% contains fats, waxes and other residual substances.

HUMUS FORMATION

Most of the humus formed in the soil results from the decomposition of plant materials, which is not possible without the intervention of microorganisms. However, the complex formation of humus needs synthesis along with decomposition of organic matter.

The organic matter decomposes into simpler compounds, which are then biochemically resynthesized within the tissues of the microbes depending on their metabolic activities, which also finally end in humus formation.

Lignin also breaks down, but extremely slowly, forming phenols and quinones. These simpler compounds are known as monomers, which later polymerize and form polyphenols and polyquinones. These polymers then combine with amino compounds and give rise to resistant humus with higher molecular weights. Their formation becomes easier in heavy soil in the presence of colloidal clay. Generally, the humic groups (polyphenols and polyquinones) are more complex which always interact with the non-humic groups (polysaccharides). The other complex group, which is known as polynurides, which synthesizes only in the tissues of microbes, may also combine with the organic portions of the soil resulting into Humus.

Out of 100 units of organic residues entering into the soil, only 15 to 30 units are changed to Humus. The rest 60 to 80 units are converted into CO_2 and are lost in the ecosphere excluding approximately 6 to 15 units, which are recycled as soil organisms plus polysaccharides and polynurides.

A simple sketch of Humus formation in the soil system from plant materials is shown in Fig. 5-1.

Humus being almost the end product of organic matter decomposition has various advantages. However, in the complex soil system it is difficult to separate between the advantages derived from humus and the other organic portions in the soil. Both have their own distinct roles. Humus is a large molecule with slow changes, but the fresh organic matter added to the soil with a rapid mode of change may have distinctly different approach in conserving soil. Humus may be able to adsorb more cations than the rest of the organic matter, which is naturally added to the soil. It is clear from these processes that the organic matter content in a soil system is naturally maintained at a certain level. With additional organic matter the level can be raised with time but it is very difficult to minimize it to a zero level of carbon.

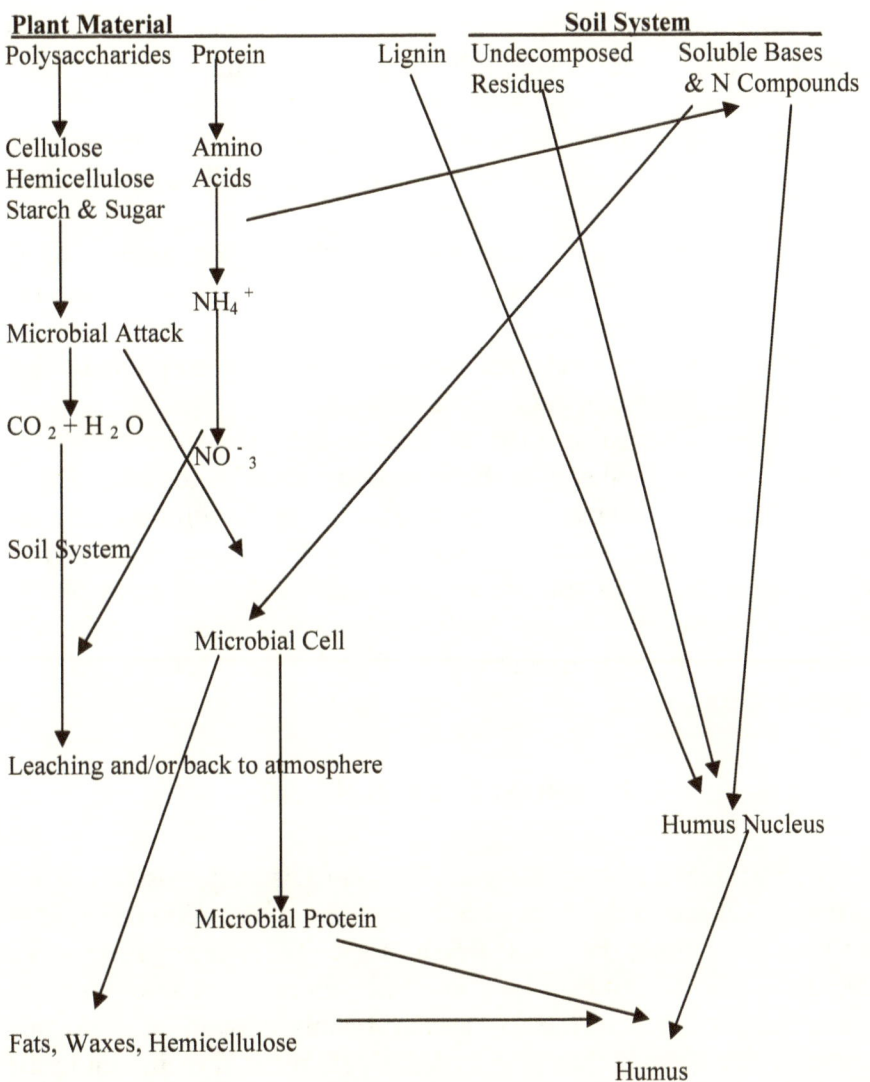

**Figure 5-1: Humus formation in soil through complex interaction
of soil and plants through microbial interventions**

The combined effects of organic matter and humus are of great practical value. The following are accomplished by their presence in soil:

1. It makes the soil color darker, which can help in retaining the solar energy in soil system more effectively.
2. Physical properties of granulation and water-holding capacity are enhanced whereas plasticity and cohesion are decreased.
3. Cation exchange capacity and the adsorption of and release of cations are enhanced.
4. They also help to release the plant nutrients from mineral soil by the increased microbial activities.

Thus, the physical and chemical activities in soil become very conducive for plant growth as affected by the presence of carbon whose main source in soil are organic matters derived from plants. However, there are some limitations, which may restrict the availability of major nutrients like S, P and N as it relates to the C contents of soil.

CARBON: SULFUR RATIO

Sulfur availability for growing plants is only possible when sulfur is oxidized in the form of $SO_4^=$ ion. Such oxidation of sulfur is not possible without the involvement of specific microorganisms like "Desulfovibrio", "Thiobacillus", and "Clostridium." These microbes attack the pool of S of the soil system mainly derived from organic matter or when applied as S -powder which helps in maintaining the C:S ratio below 200:1 for its mineralization. Such a situation can be easily achieved either by fresh addition of organic residues with low C: S ratio of 50:1, or by adding S, or by doing both.

Thus the entire activity of mineralization of S can be enhanced and maintained adequately for economic plant growth under acid soil condition especially in the biologically controlled rhizosphere.

CARBON: PHOSPHORUS RATIO

Similarly 200:1 or lower ratio of carbon to phosphorus is applicable in the case of phosphorus for its availability to plants. However, there is a major difference of soil pH, which needs to be maintained between 6.5-to 7.5-pH range for its optimum availability. Phosphorus is a very vital and key element for living organisms (microbes, plants and animal) as it is an intricate part of the genetic material of viable cells. Nitrogen application has also been found to help in the mineralization process of P, provided the P concentration is not less than 0.2%.

CARBON: NITROGEN RATIO

The major ratios: (1) C:N ratio of 17:1 and (2) 33:1 have been identified to be major markers in the soil system. These markers control the dominating process occurring under different stages as detailed below:

(a) When C:N ratio is below 17:1, N mineralization far more exceeds the process of immobilization.
(b) When C:N ratio is between 17:1 to 33:1, the process of N mineralization at that stage is equal to immobilization of N.
(c) When C:N ratio exceeds the ratio of 33:1, the process of immobilization tends to exceed the process of mineralization.
Mineralization of nitrogen simply means the transformation of organic N to inorganic N, as shown below.

For example:
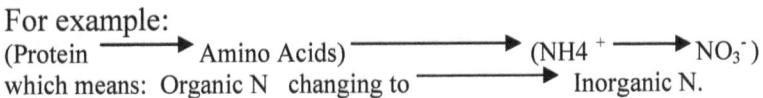
(Protein ──────► Amino Acids) ────────────► ($NH4^+$ ──────► NO_3^-)
which means: Organic N changing to ────────► Inorganic N.

Reverse process to this will be the process of immobilization. Thus when the process of nitrogen mineralization dominates, nitrogen is made available to plants.

These processes have quite a significant practical application in case of nitrogen conservation as well as its utilization by growing plants when necessary. Indirectly, it also shows the competition between the microbes and plants while growing. For the best productive level of any soil, the microbes of the rhizosphere and the growth of plants may utilize the Nitrogen alternatively. Thereby, under the increased level of organic matter, the microbes may multiply rapidly in numbers with increasing food source of carbon. On the other hand, when the carbon level goes down, the microbes decline and put less competition for nitrogen-uptake for themselves and the mineralized nitrogen becomes easily available for the growing plants. This cycle is the key for understanding the soil management processes for better plant growth. When C:N ratios are carefully managed by incorporating the right material on right time, the availability of N in plants can result into economic plant production.

The following materials need to be identified by their C:N ratios, which can be incorporated in soil system as desired to get the desired effects. Hence the C:N ratios of a few materials which go into soil system need to be known.

Table 5-1 : The C:N ratio values of materials, which enter the soil system

Materials	C:N Ratio
1. Leguminous Plants	20:1 to 30:1
2. Non-leguminous Plants	30:1 to 80:1
3. Farm Manure	90:1
4. Saw Dust	400:1

Under the adequate supply of carbon, the significance of these ratios (C:S, C:P, and C:N) as previously discussed may be summarized as follows:

1. With the fresh input of the crop residues, a portion of the added nitrogen, phosphorus, and sulfur will always be immobilized.

2. Continued cropping without adding N, S and P will lead to quick depletion of these elements due to rapid mineralization.
3. With inadequate supply of N, S, and P, the synthesis of organic matter will be curtailed.

Hence, the organic matter input and its decomposition as well as its synthesis must be a continuous cycle within a soil system for adequate plant growth.

CARBON CYCLE

Humus and Carbon dioxide are the two stable components of this cycle. CO_2 of the atmosphere enters the soil surface through the plants which are released back to the atmosphere through the living animals, soil reactions, as well as by the effective decomposition of organic matter accelerated by the soil microbes as shown in Fig. 5-2:

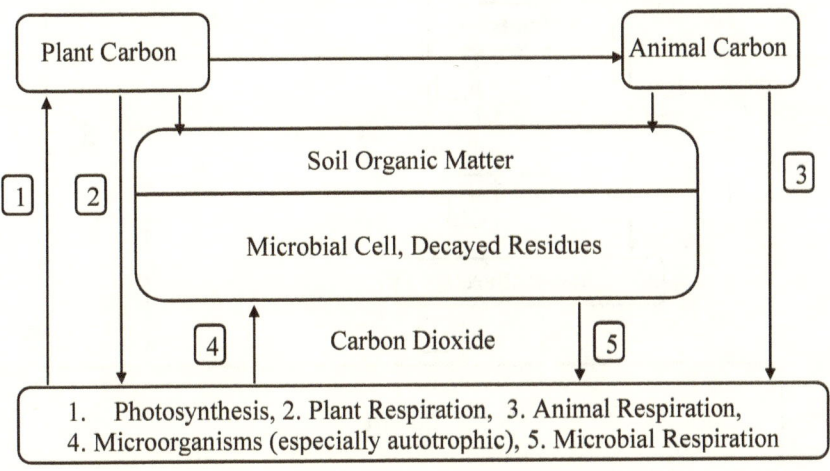

Fig. 5-2 :Carbon cycle in soil

SULFUR CYCLE

Generally half to 3/4th of total sulfur present in the soil system is in organic forms. But the entire vegetations of the earth's surface obtain their sulfur mostly in the form of $SO_4^=$ ion from soil and very little in gaseous form from the atmosphere as apparent among the industrialized nations. Animals, on the other hand, meet their demands for this element through plants. When plants or animals decay in the soil system, the earth microflora convert them to simpler amino acids, and in aerated condition, sulfur is changed to sulfate to be absorbed by plants. But in reduced waterlogged situation, the sulfur escapes as hydrogen sulfide to the atmosphere. Thus soil microorganisms help in the assimilation of sulfur in microbial cell as well as in the breaking down of complex organic substances into simpler inorganic forms of sulfur. A simple sketch of sulfur cycle is given in Fig. 5-3.

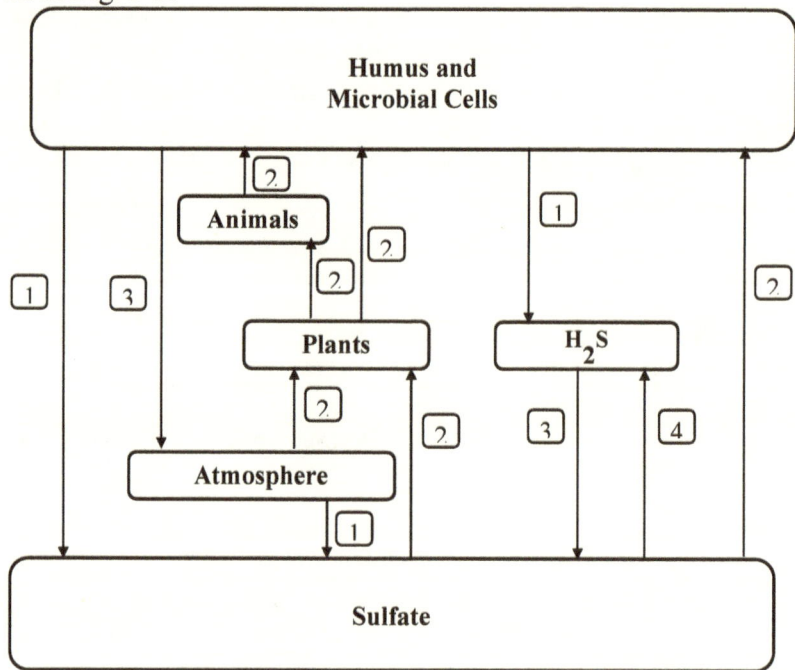

1. Mineralization 2. Immobilization 3. Sulfur Oxidation 4. Sulfur Reduction

Fig. 5-3 : Sulfur cycle in soil system

This indicates that microbes along with oxidation and reduction reaction in the soil system help in recycling of the sulfur with the help of plant life.

NITROGEN CYCLE

Out of all the major nutrients, nitrogen is one of the most important nutrients as well as the most inert element in its natural gaseous form. This passes very repeatedly from soil into the living organisms and back again into the cycle through the atmosphere. However, a very minute portion of this element may leach out and may be lost from this vital cycle.

A host of microorganism, play an important role in completing this cycle which has two major components of mineralization and immobilization within the "soil - microbes and plant system." This is one of the few soil nutrients, which, like sulfur, can be lost by leaching as well as by volatilization. But its' importance in crop production is the most critical of other essential elements known so far.

There are various microbes, and various processes, which are involved in nitrogen transformation in soil as noted below.

Name of the process	Actual Mechanism
1. Nitrogen fixation ⟶	1. Natural lightning in the atmosphere 2. By legume bacteria (Rhizobium) 3. By non-legume bacteria (Azotobacter)
2. Ammonifixation ⟶	1. By hosts of microbes (Pseudomonas, Clostridium, Fungi and Micrococcus)
3. Nitrification ⟶	1. By microbes (Nitrosomonas, Nitrosoccocus) and Nitrobacter
4. Denitrification ⟶	1. By microbes (Pseudomonas, Thiobacillus and Vibrios)
5. Nitrate reduction ⟶	1. Under reduced condition, due to poor aeration in the soil system, which can be accelerated under alkaline condition

Activities concerning nitrogen can be summarized in the Figure 5-4.

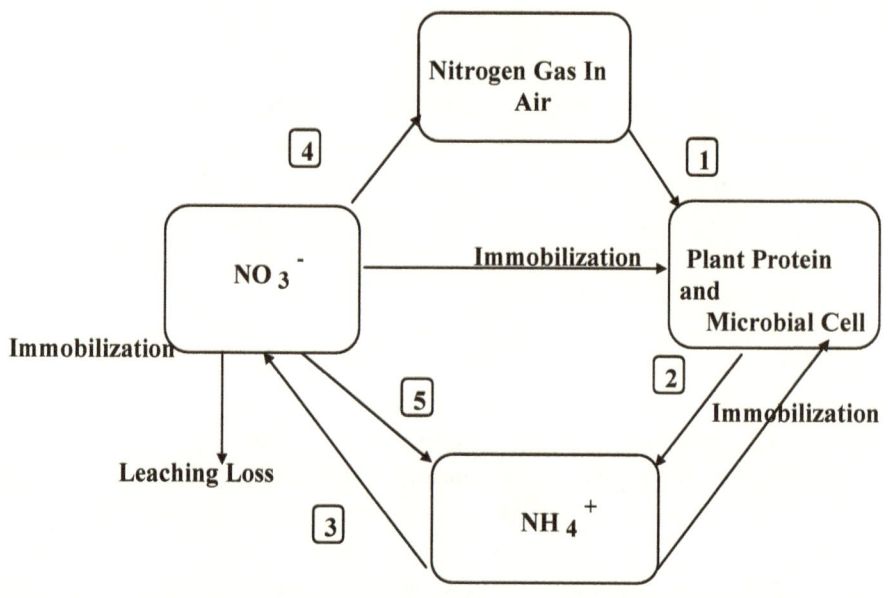

Figure 5-4 : Nitrogen cycle in soil system

SOIL ORGANIC MATTER

It is very apparent that the key elements of plant growth (C, N, P and S) have their efficient sources in the organic portion of the soil. In addition, the soil organic matter (OM) contributes to the formation, and stability of soil aggregates which next to soil texture is a good indicator of the status regarding degree of soil erodibility (the detachability and removal of soil particles). Table 5-2 confirms this.

Table 5=2: The erodibility factor K for soil showing its increase due to drop in organic matter at varying sand contents under similar rain impacts creating constant erosivity (R)

Sand (%)	OM (%)	Erodibility (K)	Erosivity (R)	Soil Loss (Tons/Acre/Year)
5	3	0.28	100	28
5	1	0.36	100	36
20	3	0.35	100	35
20	1	0.47	100	47
30	3	0.40	100	40
30	1	0.53	100	53

In the above table, erodibility refers to the property of soil to erode and erosivity refers to the capability of water to cause erosion. Hence, erodibility and soil loss are closely related.

THE PRESENT TREND OF ORGANIC FARMING

At the outset, it should be clearly understood, how the organic farming system differs from the conventional farming system. At present, the conventional system of farming includes the use of chemical fertilizers and pesticides, which are the main inputs. On the other hand, organic farming system largely excludes the use of chemical fertilizers and pesticides. Thus the organic farming system heavily relies on: (a) crop rotation, (b) manuring, (c) mechanical cultivation as needed, (d) organic fertilizers, (e) minimum tillage, and (f) biological pest control.

Various results have confirmed that farmyard manure increases the micropores in soil, which indirectly enhance the efficient use of water available to plants by encouraging the water-holding capacity of the soil. Although such agronomic benefits due to improved physical properties may be small, they will be increasingly important as crop yields approach more closely to their potential maximum.

The technique for maintaining organic matter levels used by organic farmers are essentially traditionally husbandry methods; but the function of organic matter in soil provides valuable knowledge applicable to any agricultural system. Both the practice of fertility renewal and the knowledge of organic matter functions are for the organic farmer nothing but a biological approach to agriculture. These techniques appear to work, and both production in the field and the research results, tend to support the values of this biological and organic approach. It has also been indicated that if the pH of soil is more than 7, pH no longer tends to decrease with the addition of farm yard manure, but if the pH is less than 6 it tends to increase. Knowing the pH requirement of most of the agronomical plants being close to 6.8, this appears to be very beneficial.

The properties of organic matter are utilized by organic farmers largely on a practical basis and the fine tuning of the system can often produce results, which are as profitable and productive as the conventional counterpart. However, the produce may be of higher quality and in terms of soil conservation, it would be a more sustainable system.

With the improvement in crop varieties and mechanization, the conventional farming, which has developed over the past fifty years, is largely based on the increased input of agricultural chemicals, fertilizers, and pesticides. These developments have been based on the modern theories of plant nutrition, which is linked to the use of a range of simple ionic nutrients alone. However, over 150 years ago, the humus theory suggested that the plants were nourished by the direct uptake of humic nutrients from the soil. This was apparently disapproved by the work of a famous soil chemist Liebig and others and was replaced by mineral theory. The modern theory, as such, leaves little place for organic matter other than improvement of soil's structure plus its nutrient and water supply and say nothing much about the importance of humic substances in plant nutrition. As such, the role of organic enzymes has totally been ignored. Thus there seems to be a dichotomy in modern and conventional agriculture. On the one hand, there is a theory and practice of crop production based upon knowledgeable soil management plus the use of soluble nutrient ions; on the other hand, there are extensive data, which suggest that the soil organic matter is more directly involved in plant nutrition and crop production than just as a source of nutrient ions. The successful organic farmers are demonstrating the latter to be so, on practical basis. The coming years in agriculture, with more emphasis on environment and quality products will demonstrate this more and more.

BIOLOGICAL SYSTEM

Addressing the following activities can very well summarize this system in soil:

I.   Sources of organic matter in soil
II. Factors affecting their accumulation in soil
III. Main functions of organic matter
IV. Obvious impacts of organic matter as they decompose

I. SOURCES OF ORGANIC MATTER IN SOIL

1. Plant roots get stronger with the addition of phosphorus and ultimately store "lignin" which finally adds to the soil organic matter.
2. Crop residues, which are ultimately left in the field after harvests, add to the soil organic matter.
3. Green manuring, when undertaken with legumes in the crop rotation, helps in increasing the soil organic matter as well as nitrogen, which are easily decomposed to add to the existing organic matter in the soil.
4. In the long run, stable manure is a very valuable commodity; which adds to soil's productivity by adding organic matter and especially the nitrogen to the soil system.

II. Factors Affecting the Accumulation of Soil Organic Matter

1. <u>Climate</u> - Jenny (1941) has shown that temperature is more effective in the accumulation of organic matters in soil. With lower temperatures, the Canadian soil may have 9% organic matter whereas the warmer climate of Southeastern USA may have only less than 2% organic matter.

2. <u>Soil Moisture</u> - Humid soil has 2% more organic matter than sub humid soil.

3. <u>Local Factors;</u>

a). Grasses give rise to more accumulation of organic matter than forest because of quick recycling. However, this difference is minimized at 70^0F as compared to that at 45^0F.

b). Drainage: In well drained soil (well aerated soil), the organic matter accumulation will be slow, because of its rapid decomposition due to aeration. This phenomenon ultimately shows up in soil's color.

> (b1) A well drained upland soil generally shows light color.
> (b2) An intermediate slope with moderate drainage shows medium color.
> (b3) Low lands with poorly drained situation show dark color.

4. <u>Nutrients</u>

With adequate supply of nutrients, more yields of grains, more roots and more biomass are formed which are turned to the soil. In long run roots contribution for organic mass is greater than that of the upper biomass of plants in cultivated areas.

5. Texture

The fine textured soil will have double the organic matter than the coarse textured soil, other things remaining constant.

6. Slope Aspect and Locations

The soils on North slopes contain more organic matter than soils on South slopes in Western Hemisphere.

Higher altitudes have lower temperature and hence have more accumulation of organic matter.

III. MAIN FUNCTIONS OF ORGANIC MATTER

1. Aids in water management:
> (a) Causes soil aggregation, which makes soil porous and increases water use efficiency.
> (b) Drainage in fine textured soil and water retention in light textured soil is improved.

2. Food for soil organisms:
> It is the habitat of microflora. The CO_2 given off by the microorganisms and the production of certain organic acids increase the capacity of soil liquids to dissolve the minerals to be taken in by higher plants.

3. Increases C.E.C. of soil:
> OM serves as a store house of exchangeable nutrients and a larger quantity of plant food can be applied at onetime. Next to photosynthesis, C.E.C. is important in agricultural production, which is enhanced with organic matter additions.

4. Minimum leaching loss:
> Due to adsorption and absorption of the ions, loss of nutrients by leaching is minimized.

5. A nutrient source:
> 1. Practically all the N in soil is in the form of organic matter.
> 2. P and S also increase with increased organic matter inputs.
> 3. Mineralization is increased by liming, which helps in the increase of microflora and subsequent immobilization
> 4. Stabilizes soil structure: Organic matter is responsible for aggregate formation in soil.

IV. OBVIOUS IMPACTS OF ORGANIC MATTERS AS THEY DECOMPOSE

> 1. Plant growth is faster due to effective release of plant nutrients.

2. Crop yield is higher.

3. Soil formation is expedited by microbial and chemical activities.

4. Soil structure is improved due to this breakdown, which helps in humus formation.

5. Makes the soil porous.

6. Makes the soil microbes decline which releases more plant food, as mineralization exceeds immobilization.

BIBLIOGRAPHY

1) Allison, Franklin E., 1957, Nitrogen and Soil Fertility. p. 85-94. Soil Yearbook of Agriculture.

2) Bartholomew, W. V., 1965, Mineralization and immobilization of nitrogen in the decomposition of plant and animal residues, p. 285-306. In W. V. Bartholomew and F.E. Clark (ed.), Soil nitrogen. Agronomy Monograph No. 10. American Society of Agronomy, Madison, WI.

3) Broadbent, F. E., 1957, Organic Matter. p. 151-156. Soil Yearbook of Agriculture.

4) Jenny Hans, 1941, Factor of Soil Formation. A Classic of the Soil's Literature. McGraw-Hill. New York.

5) Lugo-Lopez, M. A. and J. Juarez, Jr., 1959, Evaluation of the effects of organic matter and other soil characteristics upon the aggregate stability of some tropical soils. J. Agricultural Univ. Puerto Rico 43:268-272.

6) Martin, Alexander, 1961, The Carbon Cycle, The Nitrogen Cycle and Microbial transformation of Sulfur. p. 115-203, 225-305 and 350-367 respectively. Soil Microbiology 2nd edition.

7) Nyle, C. Brady, 1990, Soil organic matter and organic soil. p. 279-315. The Nature and Properties of Soils 10th Edition.

8) Olsen, S. R., and W. D. Kamper, 1968, Movement of nutrients to plant roots. Adv. Agron. 20:92-151.

9) Sanchez, P. A., 1976, Soil Organic Matter. p. 162-183. In properties and Management of Soils in the Tropics. ed. P. Sanchez.

10) Shreiner, O., and B. E. Brown, 1938, Soil nitrogen. p. 361-376. In Soils and Men. The yearbook of agriculture. U.S. Dept. of Agri. U.S. Government Printing Office, Washington, D.C.

11) Tisdale, J. M. and J. M. Oades, 1967, Organic matter and Water soluble aggregates in soils. J. of Soil Science 33:141-163.

12) Tisdale, Nelson and Beaten, 1985, Soil and fertilizer nitrogen, soil and fertilizer phosphorus, Soil and fertilizer sulfur. p. 117-148, 200-232, 292-304. Soil Fertility and Fertilizers 4th edition.

13) Woodruff, C. M., 1950, Estimating the nitrogen delivery of soil from the organic matter determination as reflected by Sanborn field. Soil Science Society of America, Proceedings, 14: 208-212.

CHAPTER 6

Soil Survey and Classification

SOIL SURVEY AND MAPPING UNITS

Soil is an important natural resource on the land surface. In nature it exists under varying climatic and management systems. An individual soil at any location varies more instantly as we go vertically deeper into the soil as compared to its variation as we move horizontally across the landscape. The horizontal variations at times are so diffused and gradual that it makes difficult to delineate one soil from the other without probing vertically into the soil system. Therefore, to delineate one soil from the other, we use a base map (either a topographical sheet or an aerial photo map) and study the soil profiles at intervals followed by probing the soil in between the profiles to pinpoint the changes if any. Such field study of soil profiles followed by laboratory studies, which helps in the delineation of existing soils in the area is known as soil survey. Study of a soil profile consists of the following careful examinations of each soil layers:

1. soil depth,
2. genetical layers (O, A, E, B, C and R horizons) and their definite measurements including the distinguished layers of horizons,
3. soil color,
4. soil pH and other soil chemical properties,
5. soil texture,
6. soil structure,
7. root depths,
8. concretions, and
9. certain special features at any particular depth within the profile.

These special features may include any kind of pans (clay pans, fragipans, plow pans) root concentrations, soil moisture regimes, concretions and insect holes in their natural surroundings. A trained person in soil survey must have the ability to differentiate the horizons O, A, E, B, C and R based on the following defined criteria:

O Horizon: An organic horizon composed primarily of recognizable organic materials in various stages of decomposition.

A Horizon: The surface horizon composed of various proportions of mineral materials and organic components decomposed beyond recognition.

E Horizon: Zone of eluviations resulting from intense leaching, which is characterized by a gray or light grayish brown coloration.

B Horizon: Zone of illuviation. The horizon enriched with minerals, organic materials and carbonates leached from the A or E horizons.

C Horizon: Horizon characterized by the unweathered minerals, which is known as parent material from which the soil above has been formed.

R Horizon: Bedrock.

Soil survey helps us in increasing our understanding about soil whose main objective, in general, is to have an improved utilization of land. Besides, the knowledge gained at one place concerning its use can be easily transferred to other locations provided the soil properties match. This is one of the greatest advantages of soil survey. Thus the research findings of one location can be transferred to other similar locations. According to the needs, soil surveys are generally grouped into five kinds which range from very detailed mapping to a generalized mapping depending on the mapping scale, minimum size delineation, and the kind of components used for mapping units. This is represented in Table 6-1.

Table 6-1. Criteria describing kinds of soil survey for identifications

Kinds of Soil Survey	Mapping Scale Size	Minimum delineated Size	Components Used in Mapping Units	Types of Planning Desired
1st order	<1:12,000	<1.5 acre	Phases of Soil Series	Detailed to very Intensive
2nd order	1:12,000 to 1:31,680	1.5 acres to 10 acres	Phases of Soil Series	Detailed
3rd order	1:24000 to 1:250,000	6 acres to 40 acres	Phase of Soil Series and families	General to Specialized
4th order	1:100,000 to 1:300,000	40 acres to to 625 acres	Phases of Soil families and sub group	General
5th order	1:250,000 to 1:100,000	625 acres to 10,000 acres	Phases of sub groups, great groups, sub-order and order	General to very broad

SOIL TAXONOMY AND CLASSIFICATION

The presentation of such soil data as indicated on Table 6-1 has a spatial concept of representation resulting into a soil map. However, there exists another form of soil representation, which is more narrative and tabular. This is identified as the vertical scheme. These are the taxonomic units of soil classification. Soil taxonomy uses organized sets of soil's properties and puts them into corresponding taxonomic classes, which truthfully do not occupy space, as it is a vertical system. On the other hand, the mapping unit is a horizontal system occupying space in the landscape. The

107

mapping unit is based on the concept of pedon (a three dimensional smallest unit of an individual soil) that occupies space in the landscape. The accuracy of describing the pedons in detail may result into the accuracy of soil delineations and may considerably improve the management practices of the soil. The soil mapping unit is flexible. It can be manipulated to meet the needs of different intensities of soil surveys. In detailed surveys, delineations are small and the mapping units' components are precisely defined. As the survey becomes more generalized and less intensive, soil delineations are larger and mapping units are broader and a bit vague. In such cases, mapping units may have two or more soil taxonomic units. The behavior of the soil mapping units may be based on the interaction between the two soil taxonomic units and the accuracy of delineations may be lost. Up to the period of C.F. Marbut, the soil survey guide lines were based on the mapping units formulated on broader soil qualities as compared to the specific considerations of the quantification of soil's properties which gave rise to the modern Taxonomic classification of soil. This system of classification is very scientific which has the most acceptability for soil as a very dynamic and vibrant natural body, which is always in a mode of change. This system scientifically resembles the classification of elements as arranged under the periodic table where all the elements have been separated on certain specific quantitative criteria. Even the newly discovered elements in coming future will have their clearly defined place in the table. This is possible under the modern technique of Soil Taxonomy by which any soil anywhere at anytime can be classified and scientifically placed and named.

There is a clear-cut difference in the land classification and soil classification. Land classification is based on the significant physical and cultural characteristics, whereas soil classification is based on soil characteristics alone. It does not consider economic and social factors, which are considered in land classification system.

DIAGNOSTIC HORIZON IN SOIL TAXONOMY

The diagnostic horizons are used to classify soil into various orders. They may be considered into two major groups:
(1) diagnostic surface horizons known as epipedons, and
(2) diagnostic sub-surface horizons.

Diagnostic surface horizons or epipedons: The following six horizons are identified under this group.

1. Mollic,
2. Anthropic,
3. Plaggen
4. Umbric,
5. Ochric, and
6. Histic.

1. Mollic epipedon: This epipedon can be identified by the following properties: (a) the surface soil contains less than 250 part per million of P_2O_5 soluble in 1 percent citric acid. (b)Dark color soil is having value of 3.5 when moist and 5.5 when dry with chroma less than 3.5 when moist. (c) Base saturation is 50% or more. (d) The upper mineral soil layer contains more than 4% organic matter with minimum thickness of 10 cm with soft feeling when walked on it.

2. Anthropic epipedon: Like mollic but having less than 50% base saturation with more than 250 ppm of P_2O_5 content due to human remains or human activities.

3. Plaggen epipedon: This is produced by human activities caused by long and continuous manuring. The manure from the animals with sod used for bedding must have produced this on the cultivated land having thickness of 50 cm or more.

4. Umbric epipedon: Thick dark colored epipedon very high in organic matter with base saturation lower than 50%. However, it lacks the softness of mollic epipedon when walked on it.

5. Ochric epipedon: Light colored; thin to be mollic, anthropic, and umbric or plaggen; and common in youthful soil under forest vegetation.

6. Histic epipedon: Having organic horizon, over 20 cm thick. The organic carbon in this layer may be 12% or more in light soil and 18% or more in heavy soil textures.

Figure 6-1 indicates the relationships of these epipedons to comprehend.

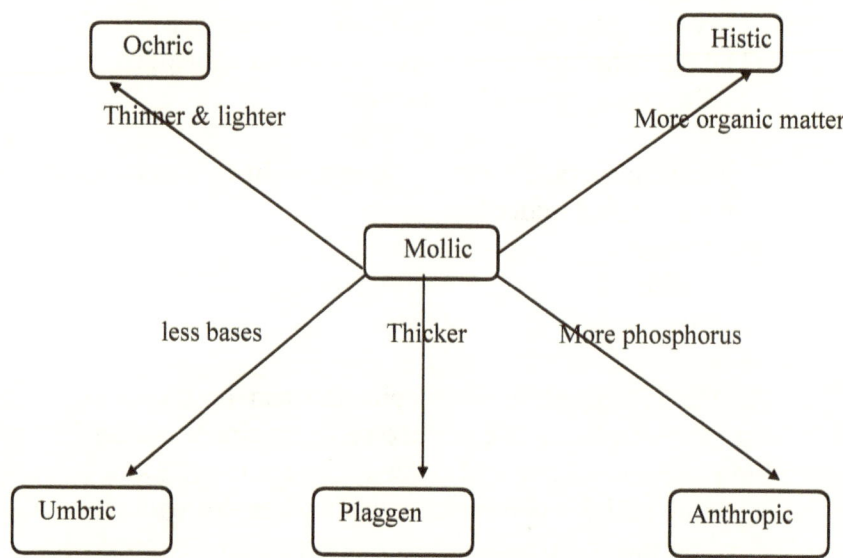

Figure 6-1 : Epipedons as they relate to mollic epipedon.

Diagnostic subsurface horizons: six such sub horizons have been identified:

(1) cambic, (2) argillic, (3) natric, (4) spodic, (5) oxic, and (6) agric.

(1) Cambic subhorizon: This horizon has features that represent genetic soil development but without mineral accumulation by illuviation and weathering. Clay contents do not show an increase with depth. No clay skins are found.

(2) Argillic subhorizon: Where clay shows an increase of 20% as compared to the upper epipedons. This shows the sign of clay illuviation, which has eluviated from the upper layers of soil.

(3) Natric subhorizon: This subhorizon along with argillic horizon may show 15% increase in the saturation of Na^+ as compared to the upper epipedon.

(4) Spodic subhorizon: This is also an illuvial subsurface horizon but with thin A2 albic horizon under which this illuviated horizon rich in iron and aluminum oxides with organic matter forms.

(5) Oxic subhorizon: It is a 30 cm or thicker subsurface horizon, which dominates the hydrated oxides of iron and aluminum with Kaolinite. These oxides of iron and aluminum are the indicators of highly oxidized subhorizons known as oxic.

(6) Agric subhorizon: This has formed due to many years of cultivation. Where the upper layers above the plow layers have changed as compared to this subsurface agric horizon due to long-term cultivation, which is in contrast to the soil structure above. These agric horizons are deposited as lamellae.

All these diagnostic horizons help in the understanding of the soil classification principles in soil taxonomy, which has resulted in developing the various categories of soil.

SOIL CATEGORIES AND THEIR NAMES

It has six categories or levels within the system. They are order, suborder, great group, sub group, family, and series. In the United States alone, there are at present eleven soil orders and more than seventeen thousand series already established.

The categories of soil are schematically presented in Fig. 6-2.

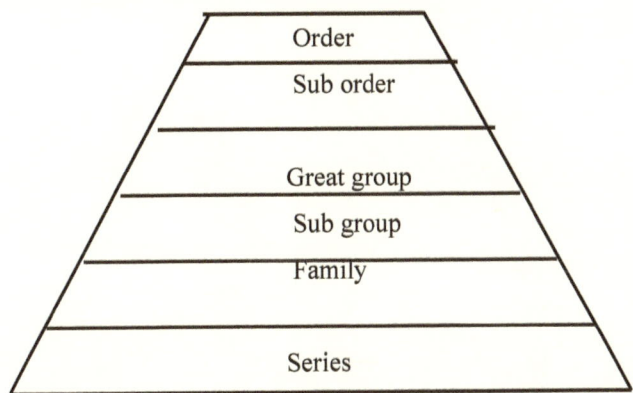

Fig. 6-2 : The hierarchy of soil levels or categories

Soil orders are taxonomically defined on the basis of soil characteristics based on the following:

1. Degree of Horizonation
2. Index of weathering
3. Diagnostic Horizon and
4. Gross Composition of the soil profile

Suborders are differentiated on the basis of (a) moisture regime, (b) temperature regime, (c) chemical and physical properties differentiating the parent material and the effect of soil forming process. Great group has the further sub division of suborder which emphasizes (a) base status, (b) presence or absence of diagnostic layers, and (c) the degree of expression of horizons.

Family is defined mainly by the texture, mineralogy and temperature regime of the sub soil horizons and/or the control zone. Soil series are based on the name of the place where the pedon of that character was first identified and characterized based on the general setting of the genetical layers irrespective of the top layer.

Classifying a soil needs morphological description of a soil profile and the systematic steps to determine soil order, diagnostic horizon, soil moisture regime, and following the keys to soil taxonomy for determining the lower categories. These may need the back-up of the laboratory analysis too.

Table 6-2. Orders and their distinguished characteristics

Order	Short Descriptions	Diagnostic Horizon	Diagnostic Features
1. Entisols	Very weakly developed soils, just starting to form, B horizon absent	None	It is of recent origin
2. Vertisols	Soils with high shrinking and swelling properties	Having rough microrelief at the surface (gilgai)	Shrinking & swelling clays dominate
3. Iceptisols	Weakly developed soil Slightly B horizon formed	Thin B horizon and may have cambic horizon	Little Profile Development
4. Aridisols	Soils of arid environment	None	May have ochric or agrillic but no oxic or spodic
5. Mollisol	Soil with thick, dark surface horizons which are high in organic matter content	Mollic epipedon	May have argillic but not oxic or spodic
6. Spodosol	Just below the bleached horizon, there is accumulation of aluminum, iron and organic matter	Spodic or highly Bleached sub surface horizon	No oxic or argillic horizon
7. Alfisol	Soil with sub soil accumulation of silicate clay having a high base saturation.	Argillic horizon	May not have mollic, oxic, or spodic content
8. Ultisol	Low base saturated soil and highly weathered	Argillic horizon	May not have spodic or oxic
9. Oxisol	Highly weathered soil rich in oxides of Fe and Al.	Oxic horizon	May not have spodic or argillic horizon

| 10. Histosol | Soils formed from organic material. | None | Organic soil with no mineral diagnostic horizon |
| 11. Andisol | Soils formed from volcanic material | — | Andic material |

Excepting the Histosols and Andisols, all the other orders from Entisol to Oxisol (1 to 9) maybe arranged in the same order showing their degree of developments as indicated by the graph in Fig. 6-2. This represents, Entisols and Oxisols as the least developed and highly developed soil profile respectively, other orders being in between these two.

Also, it may be necessary to make clear that the horizonation activities gradually fade and become indistinct as we move either to the recently formed soil 'Entisols' or to the very highly developed soil 'Oxisol' having usually oxic horizon. This is indicated in Figure 6-3.

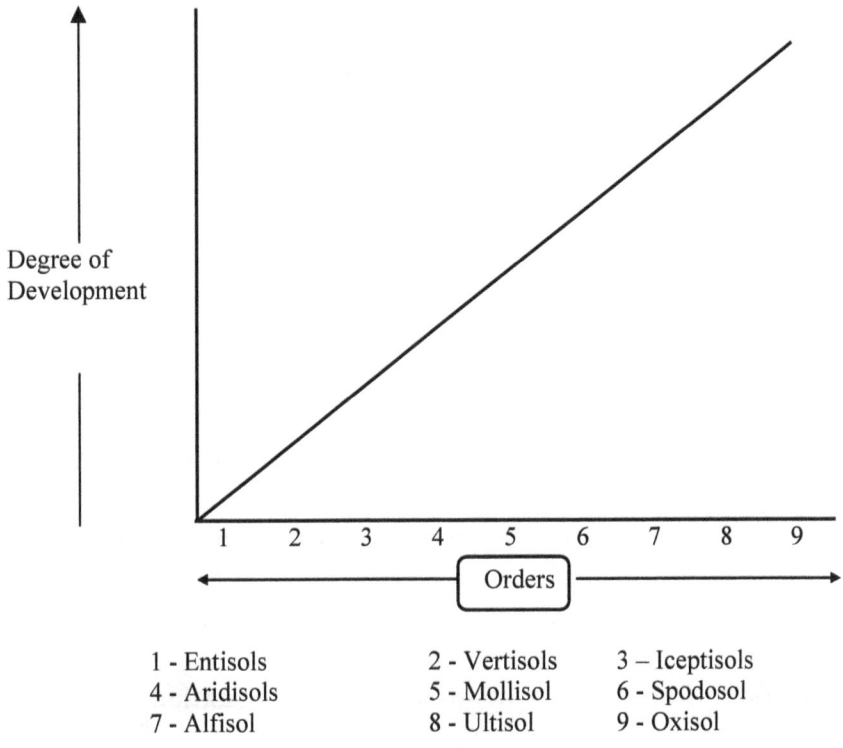

1 - Entisols   2 - Vertisols   3 – Iceptisols
4 - Aridisols   5 - Mollisol   6 - Spodosol
7 - Alfisol   8 - Ultisol   9 - Oxisol

Fig. 6-2. Comparative developments of soil orders

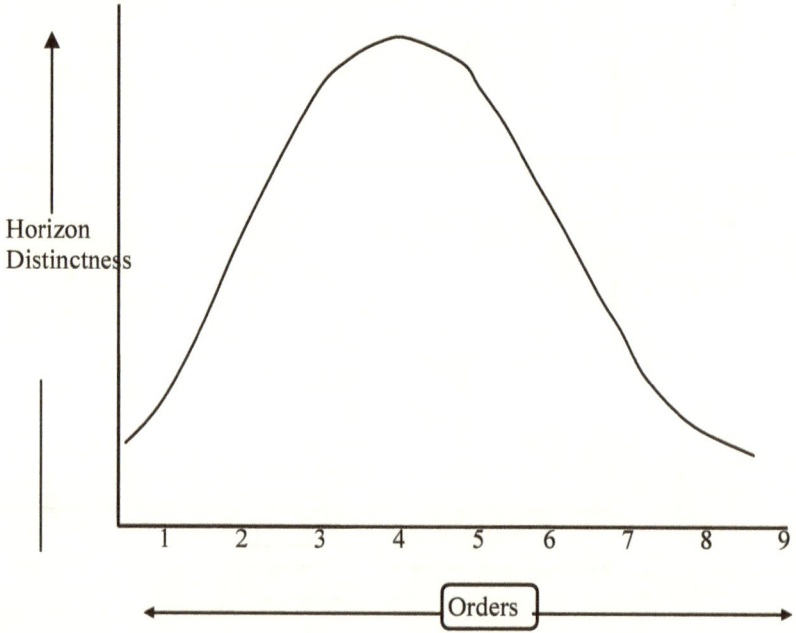

Fig. 6-3 : Comparative distinctness of horizonation of soil orders

The next step in classification is to determine the soil moisture regime like: udic (humid climate); ustic (semi arid climate); xeric (mediterranean climate) and aridic (arid climate of the desert). There is another one known as aquic, which denotes the local factor like topography and parent material, which may cause higher soil water table with low conductivity of water.

Following similar steps throughout the lower categories up to family, the names of the examined soil can be given. For example, the full name and its components are explained below.:

Full Name:	Typic Hapludalf silty, mixed, thermic
a	'alf'
b	'udalf'
c	'Hapludalf'
d	'Typic Hapludalf'
e	'silty, mixed, thermic'

The explanation of the above is systematically noted:

a – 'alf ' denotes Soil Order "Alfisol."

b – 'udalf ' denotes Sub Order developed under hot humid condition.

c – 'Hapludalf ' denotes Great Group showing weak horizonation.

d – 'Typic Hapludalf ' denotes the central concept of sub group.

e – 'Silty, mixed, thermic' denotes silty texture, mixed mineralogy and warmer climatic conditions.

At the end of this nomenclature, one can add the name of the place where such profile was examined for the first time plus the texture of the topsoil. This will appropriately complete the taxonomic classification of the soil, which for example will be "Typic Hapludalf silty, mixed, thermic Memphis silt loam." Such names will clearly indicate the nature and properties of the soil as identified and acceptable within the community of soil classifiers.

Historically soil classification was initiated in China, forty centuries ago for tax purposes. In the 20th Century, Russia became the leader, when V.V. Dokuchaiev considered soil to be an independent natural body. After that USA became the leader in organizing the soil classification system, which may accommodate the entire concept of soil quality based on Russian views.

In recent years, a change in the concept of soil has been intensified which has improved on the past views on soil. This is based on quantitative measurements of existing soil and is viewed as dynamic and accepts the unavoidable continuous change in soil, which is always possible. This change has continued and will continue in the future. The current classification system (soil taxonomy) is quantitative and is based on the following modern concepts of soil:

1. The lower limit of soil is now the lower limit of biological activity.

2 Even buried soil is soil, so long as it has biological activities.

3 Alluvial soil is soil if it can support plant life.

4 Water which supports plants is not soil but underneath water, the plant growing material is soil.

5   The aboveground parts of the plant, which support plant, are not soil.

6   Soil is not continuous as has been comprehended by Marbut.

Although basic concept of soil classification by examining soil profile continues, it is no more only qualitative. The properties have to be quantified for classifying soil in this modern era and that is exactly the case in the present trend of soil taxonomy and classification.

BIBLIOGRAPHY

1) Baldwin, M., C.E. Kellogg and J. Thorp. 1938. Soil Classification. Page 970-1001. In Soils and Men USDA Agricultural Yearbook, Washington, D.C.

2) Bartelli, Lindo J. 1979. Interpreting Soil Data. Pages 91-114. Planning The Uses and Management of land. No. 21 in the Series Agronomy. ASA, CSSA and SSSA.

3) FitzPatrick, E.A. 1971. Pedology. Oliver and Boyd, Edinburgh.

4) Olson Gerald W. 1981. Soils and the Environment. A Guide to Soil Surveys and their Applications.

5) Pepper, I.L. 1996. Abiotic Characteristics of Soil. p. 9-18. Pollution Science. Academic Press. New York.

6) Perkins, H.F., Tiwari, S.C., and K.H. Tan. 1971. Classification and Characterization of a Highly Weathered Soil of the Southern Piedmont. Soil Science Vol. 111: p.119-123.

7) Soil Survey Division Staff. 1993. Soil Survey Manual USDA Handbook #18. Washington, D.C.

8) 7. Soil Taxonomy - Achievements and Challenges. 1984. SSSA Special Publication #14. Soil Science Society of America, 677 South Segoe Road, Madison, WI 53711.

9) 8.Wilding, L.P. and K.W. Flach. 1985.Micro-pedology and Soil Taxonomy. p. 1-12. Soil Micromorphology and Soil Classification SSSA special publication #15.

CHAPTER 7

Managing Soil Environment for Crop Production

In order to fully comprehend the entire phenomena of soil formation and its significant behavior, it is mandatory to have an in-depth look into its ingredients. This facilitates the understanding of the soil in your possession while you discover its potential for raising a particular crop or series of crops. Even then, the edaphological approach for your soil may not be so easy. Because, similar soils may require different management skills depending upon their locations, the economy, varying past treatments, climatic conditions, and the specificity of the crop to be raised. It is true that all the above portion of the soils environment have their unique contribution. However, the main resource 'Soil' itself has it's special importance; when one imagines that this is the only viable resource on this planet earth which has to feed 8+ billons of people around the world. This is a huge undertaking.

In order to meet this challenge, we have only four biological bases to raise the food and fiber to support this growing population. All these four bases are directly or indirectly fueled by solar energy to photosynthesize food manufactured by selected green plants, which are made to anchor into our soil. These four bases are fisheries, forestry, grassland, and croplands.

For all the above four base areas, to lesser or grater extent, Fertilizer Technology of 19 [th] century has so far worked fine in increasing our productions. Specially, the fertilizers applications in croplands did have the highest contribution in this regard. However, recent exploitations of forest plants combined with the misuse of fertilizers particularly in croplands have caused enormous degradation of our soils. In lack of other innovations and technology to increase our food production, we are left with only alternative to take care of our soil resources to the best of our abilities. The objective of our actions must focus towards increased crop production as well as its sustainability, which depends on the up-keep of our present soil. In addition, such effort may require genetically improved crop seeds, suitable crop rotation, pest control measure, fulfilling the water need, and measures to eliminate or minimize the deterioration of our soil environment. Such things need constant vigilance and need timely adjustments. Because, soil and selected plants association is a vary dynamic system. There-fore then it is imperative to note down certain general directions which need to be followed, if we look forward and plan to use soil for crop production on a sustainable basis:

1. ACTIVE STEWARDSHIP BACKED BY RECORDS

This is needed if one wants to really take care of land for crop production now and in future. This will require a General Soil Map of the farming area indicating the input at different parcels of the land and the total harvests obtained year wise. This needs to be maintained on each parcel and on each of the crops raised with notes and remarks indicating the specific reasons of successes and failures. This record will also include the Soil Test results at certain intervals indicating the period of sampling and the previous crop and land use when soil samples were taken. This entire record generated regularly will guide you now as well as in the future indicating ways and means for improvement. Such record keeping is a key for successful farming and an excellent stewardship for passing it on to the next generation.

2. ADJUSTING OR MODIFYING CROPS ACCORDING TO THE SOIL CONDITIONS

The time has come to modify crop lands rather than attempting to change soil environment to meet our economic challenges in raising certain crops at various locations. In the past we always did the reverse and we have found it to be vary costly. In doing so we always have wasted our energy to fight against the nature. Now we must revise our views and make use of the plants genetics. Crops can be economically modified to match the existing

121

soil. Such steps taken for matching are essential for better harvest under the sustainable farming enterprise.

3. CROP ROTATION

Crop rotations are the must for maintaining longer viability of a healthy soil environment. This process directly eliminates the monoculture, which generally harbors the organisms for plant's diseases. Besides, this practice helps in using the full potential of soil by utilizing the entire rhizosphere in alternate succession. In general, the deep rooted crops need to be rotated with shallow rooted crops to accomplish the goal. From nutrient point of view, the non-legumes need to be rotated with legumes for balanced absorption and retention of nutrients by the root and soil system respectively. Soil's structures, textures, and the existing hard pans be carefully considered to make the entire concept of the crop rotation to succeed. Under this method of farming all the plant portions excepting the harvest must return to soil.

4. SOIL MOISTURE MAINTENANCE FOR CROPS

In general 80% of the plant's weight is just water. Still the water requirement of one crop always differs from the other crop. Besides, the moisture holding capacity of soils also differs from one soil to the other depending on the rhizospheric soil depths, organic matter contained, soil texture, as well as on the proportion and distribution of the micro and macro pores present. The total amount of water required by a specific crop and the critical stages in plant's life when they need most water need to be fully recognized. Because, total amount of water present in soil has the least importance as compared to the availability of water in the rhizosphere at critical stages. These stages in plants are at early growing stage, flowering stage, and at the stage of fruit set. At these three stages soil moisture must be with in the available scale range of 1/3 to 15 atmospheric tensions. Preferably 5 to 10 is consider to be the best for efficient plants absorption. Crop producer needs to understand that too much or too little water causes unusual strain on the plant and both the situation create wastage of energy to the plants and the yields are reduced.

5. Soils Slope and Slope Directions

In order to efficiently conserve soil moisture and simultaneously avoid soil erosion, cropping must always be laid dawn against the existing slope. Lands with slopes need contouring. The object of contouring should never be to fight the nature but to take advantage of the existing slope by careful manipulation to slow down the running water on the land surface. This avoids erosion and increases the span of time for the running water to permeates into the soil for storage and subsequent use by crops. The space used in between the contour lines should be kept wider in heavier soils than in the lighter textured soils depending on the roughness of the soil surface and the soil's permeability. Usually land having south-westerly slopes remain drier as compared to the fields facing north-easterly direction in the northern hemisphere. This indicates that under similar conditions soil water need may be higher under south-westerly slopes as compared to north-easterly slopes. This may help in planning the irrigation frequencies under the specific slope directions. Having similar permeability and slopes, the heavier soil may always need more intervals in between the water inputs by irrigation as compared to light textured soils. This will hold true for south-westerly as well as north-easterly slopes.

6. Soil pH and Crop Adjustments

The acceptable pH of soil for various crops range between 5 and 7. This must be examined more frequently. If possible, the crops suited to the existing pH must be preferred. Any attempt to change the naturally existing pH must be avoided if feasible.

Soils in general have the tendency of getting low in soil pH with time. This is due to the removal of the bases by crops and / or being washed away with the elapse of time. At pH lower than 5, Al^{+++} and Mn^{++} become more soluble and become more toxic to the plants. Under this condition either acid loving plants need be grown or this condition needs to be amended by liming. Within the pH 5 and

7 basic cations' stability increases, clays flocculate, soil aggregates become stable and create a condition conductive for plant growth with raise in soil quality. There appears to be a positive relationship between good soil quality and a good crop harvest. Besides, an established good soil quality with profitable harvests year after year may be a sign of sustainability. An experienced farmer and proficient agronomist always do recognize these relationships.

7. Use of Chemical Nutrients

A well managed farming system needs economical use of chemicals. Without using fertilizers, it is impossible to raise better crops from nutrient depleted farm lands. Identifying the deficiency of nutrients during the plant growth may be essential for future inputs. But to remedy that crop during the same season when detected is too late for profitable harvest during that season. One needs to mitigate the crop needs right from the initial stage of crop growth by eliminating the existing deficiencies. This requires regular soil testing to follow recommendations in a very disciplined manner. In order to make good use of previously used chemicals, one needs to incorporate the remaining stand of crops just after the harvest. This is a fairly acceptable management practice to maintain soil organic matter and to build a fertile soil capable of retaining the useful nutrients for the next crop.

8. Use of Pesticides and Herbicides

Judicious uses of herbicides and pesticides have always helped to raise better crops. Use of herbicides (pre-emergence and /or post-emergence) may help in reducing the useless competition by harmful weeds against the useful crops. One must spray the recommended pesticides (insecticides and fungicides) when identified by the crop symptoms. However, it is better to follow the recommended spray schedule to avoid the outbreak of the disease altogether. It is always better to use disease free or at least disease resistant seeds in your cropping schedules. Such genetically improved seeds may be costly, but will always be profitable by bringing quality harvests.

Raising a good crop on the lap of the mother earth requires constant vigilance and a keen experienced hand. This may need a sufficient know-how for this art of modern farming associated with academic background. A scientific analytical mind will always be helpful to evaluate decisively the action as needed. The time is long gone, when multiple plowing was considered to be the key for better harvests. Now this is the age of minimum tillage or no tillage at all, provided we care for sustainable harvests through our advanced farming system. With changing time, we need to have a changed outlook for managing our croplands for the present as well as for the future.

BIBLIOGRAPHY

1) Allison, F.E, 1973. Soil Organic matter and its role in crop production. Elsevier Scientific, New York

2) Black, A.L, 1973. Soil property changes associated with crop residue management in the wheat-fallow rotation. Soil Sci. Soc. Am.Proc.37: 943- 946

3) Brown, Laster R., C.Flavin, and S.Pestel.1991. Protecting the biological base, Saving the Planet, pp.73-82.

4) Brown, Laster R., C.Flavin, and S. Pestel.1991 Grain for Eight Billion, Saving the Planet, pp.83-96.

5) Bruce R.R., A.W.White Jr., A.W.Thomas W.M.Snyder, G.W.Langdale, and H.F. Parkins.1998. Characterization of soil-crop yield relation over a range of erosion on a landscape. Geoderma 43:99-116.

6) Edwards, W.M.1991. Soil structure processes and management. In R.Lal and F.J.Pierce (eds.). Soil Management For Sustainability. Soil and Water Conservation Society, Ankeny, IA, pp7-14.

7) Hively, Dean W. and William J.Cox.2001. Interseeding cover crops into Soyabean and subsequent corn yields. Agronomy Journal 93:308-313.

8) Jaseen.B.H.1984. A simple method for calculating decomposition and accumulation of "young" soil organic matter. Plant Soil 76:279-304

9) Krupinsky, Joseph M., Kareen L.Bailey, Marcia P.McMullen, Bruce D.Gossen, and T.Kelly Turkington

2002. Managing plant diseases risk in diversified cropping systems. Agronomy Journal 94:198-209.

10) Lala.R. 1999. Soil Quality and Food Security: The Global Prospective. Soil Quality and Soil Erosion, Soil and Water Conservation Society, Ankeny, Iowa, pp3-16.

11) Lucase, R.E., J.B.Holman and L.Connor.1997.Soil carbon dynamics and cropping practices in W.Lockertz (ed.) Agriculture and Energy. Academic Press, New York, pp.33-351.

12) National Research Council.1993. Soil and Water Quality. An Agenda For Agriculture. National Academy Press. Washington D.C.

13) Odell, R.T.1950. Measuring of productivity of soil under various environmental conditions. Agronomy Journal. 42:282-292.

14) Pierce, F.J.1991. Erosion productivity impact prediction. In R.Lala and F.J.Pierce (eds). Soil management for sustainability. Soil and Water Conservation Society, Ankeny IA, pp-35-52

15) Sanchez, Jose E., Elder A. Paul, Thomas C. Wilson, Jeffery Smeenk, and Richard R. Haewood. 2002. Agronomy Journal 94:391-396.

16) Sainju, Upendra M., Bharat P.Singh, Siddat Yaffa. 2002. Soil organic matter and tomato yield following tillage, cover cropping, and nitrogen fertilization. Agronomy Journal. 99:594-602.

17) Schnitzer, M.1991. Soil Organic Matter and Soil Quality. Tech.Bull.Agric. Canada No.1991-1E. Ottawa.pp33-49.

18) Shapiro, Charles A., David L. Hashouser, Willams L.Krenz, David P. Shelton, John F.Witkwiski, Keith J. Jarvi., Gerald W.Echtenkamp, Lisa A.Lunz, Robert D.Trerichs, Ray L..Britlinger, Mari L. Luuberstedt, Melinda Marvey Mceluskey, and Walter W.Stroup. 2001.Tillage and Management Alternative for Returning Conservation Research Program Land to Crops. Agronomy Journal: 93: 850-862.

19) Stauffer, R.S., R.Muckenhirn, and R.T.Odell, 1940. Organic Carbon, pH, and aggregation of the soil of Morrow plots as affected by types of cropping and manurial addition. Agronomy Journal 32:819- 832.

20) Taylor, H.M. and E.E. Terrel. 1982. In J. Rechigl Jr.(ed) C.R.S. Handbook of Agriculture Productivity, Vol.1.C.R.S.Press Boca Raton, FL.

21) Vorley, W.T. and D.R.Kenny. 1995.Sustainable Pest Management and Learning Organization. Leopold Center for sustainable Agriculture. Iowa Sate University. Ames.

22) Young. R.A., J.C.Zubriski, and E.B.Norman. 1960. Influence of long-term fertility management practices on chemical and physical properties of a Fargo Caly. Soil Sci.Am.J. 24:124-128.

CHAPTER 8

Soil Care

Soil is an intimate part of our ecological system extended over the vast stretch of land surfaces, which exists as an intricate part of land itself. In order to take care of soil, we need to think its setting within the framework of that system. As the ecosystem of one location changes by natural processes or by the desirable and undesirable actions of human's endeavor, we need to choose for the optimum conditions for soil based on its specific situation. Here the word "optimum" in its expression is as vague as it can be; which may differ according to an individual's thinking and projections. However, a particular setting can be optimum when it preserves the integrity, stability and the beauty of the biotic community and it is wrong when it tends otherwise. Thus to maximize the optimum appearance and usefulness of any parcel of land, within any ecological system approach, all the biotic factors (humans, animals, plants, microorganisms and soils) and their interactions with air, water and chemicals need to be considered individually but the approach of its improvement must be as a whole.

Care of soil must influence the optimization of the biotic life on the land. This approach can be local, natural or international which may fulfill the intended real objective of an ecological improvement, because ecological boundaries do not follow the county, state, or national boundaries. We know that the polluted atmosphere pollutes the lithosphere as well as the hydrosphere. In the same token, the pollution in the terrestrial system will pollute others, which are indirectly or directly attached to it. This attachment is not controlled by man-made national boundaries. Nuclear power sources, pesticides, herbicides, fertilizer and the demand for excessive energy for the extraordinary amenities of life indirectly or directly produce pollution. Extremely poor or extremely affluent societies generate pollution in different modes. This makes it essential to conserve our soil resources. This means we have to attempt to make our soil productive and maintain it for an indefinite period of time.

So far as the pollution is concerned, both the so-called developing and developed countries have to take care. The developed countries may create pollution due to their affluent nature by using unnecessary amount of fertilizers, herbicides and insecticides that may create undesirable effect on the ecosystem. Besides, they can be careless in the production of such items, which cannot be easily recycled. On the other hand, the developing countries have to watch their population pressures and educational levels of each individual, so that they may be able to think and act to keep their ecosystems viable. Both the worlds have to pay their prices and one cannot thrive at the cost of others. One cannot over or under utilize the soil. As a matter of fact, developing countries have much pressure on the soil whereas most of the developed countries have less pressure on soil for production, because they can meet their needs by other means, which developing countries cannot afford.

Understanding the individual soil anywhere will help in designing proper care of the soil. Its physical, chemical mineralogical and biological properties must be comprehended well on their individual locations to provide adequate answers to any soil-related problems. This will help in designing proper care of the soil at any particular location, when all the properties justifiably focus on its adequate classification and proper use.

As such, the biological soil properties these days must have their upper hand in solving the problem of soil care. We need to examine critically the soil system based on this property, which may play miracle in the ecological system. The microbiological properties can be a source of great improvement as well as may be the cause of the human destruction from this planet earth. That is why simple drinking water has been playing havoc throughout the world today. When the microbial environment has to do with the cleansing effect on soil by providing plant food, it is a healthy environment. However, if it is releasing toxins in place of the plant food, it is a very unhealthy soil environment, which may cause great sufferings on the planet earth. Considering this, any biological product should be well composted before using in soil system to avoid human sufferings. In spite of the fact that the land treatment is an established waste treatment technique, it has a significant role to play in America and in the world. But it should be visualized both ways. It can be either the beginning of waste disposal for the productivity of soil if done wisely or it may cause the initiation of reclamation program of soil and land if done wrongfully. Anyway we look at it, recycling and reutilization or reductions of the waste volumes are the keys for environmental conservation. This can be a defensive technique in the overall proposal of "Soil Care" but the creation of better environment must be the goal, which needs to be pursued very actively. Only then we can succeed to have requisite food, feeds and fiber productions and may be able to reduce expenditures and the adverse effects of soil use for the enhancement of agricultural production in future.

The willingness to care our soil resources is a priority. A very general question then arises as how can it be done? The living identity of soil, which we must never consider as dirt, has to be kept alive in our minds to produce abundant amount of food and fiber efficiently forever. This approach must be initiated to make us mentally alert by restoring our caring attitude towards soil for our own survival. We are aware that plants out of place are called weeds. But these so-called weeds are also indicators of soil's fertility. The locations where they grow luxuriously are certainly fertile soils. They also indicate that if left unattended, they try to take over. Where such activities are more time efficient, the soil is more fertile. Indirectly, they safeguard our soil and many lessons are to be learnt from their activities. The first lesson is that soil where they grow so fast has been undisturbed. The recent findings also indicate that unnecessary disturbance of the natural soil system must be avoided as much as possible. Useless tilling and turning of soil is not needed at all. On the other hand, a blanket of natural vegetation must always be recycled without clean removal of the thick organic mat from the soil. This is the major management key for soil's improvement. Thus the deterioration of soil can be critically examined by observing and quantifying the organic contents of soil, which are always being recycled by the organisms present within the soil's surface. The soil manager must always focus on the increase of this natural habitat of beneficial organisms to regulate and enhance our soil's productivity. Soils need to be managed with a close contact and never by the

131

findings of the testing laboratories only. Soil manager must be a keen observer of soils, which he manages and must be able to read the soil's language. To do this, he must listen to those who are always in close contact with the soil and also observe and handle the soils himself. In doing so, if he manages to increase the organic contents of soil by returning all the plant refuse and residues with minimum tillage, his grain harvest will be increasing year after year. His soil test results will also indicate an increase in the amount of available nutrients and water. However, he will have to understand that different inherent qualities of soils will always differentiate the individual soil's yield potential from one another.

BIBLIOGRAPHY

1) Beaty, Marvin T., Gary W. Petersen, Lester D. Swindale, Editors. 1979. Planning The Uses and Management of Land. ASA, CSSA, and SSSA Publication No. 21.

2) Frantric, I., and E. Parrakova. 1977. Waste Management, Problems and Their Impact on the Environment. p. 797-806, Proceedings of the 1976 Cornell Agricultural Waste Management Alternative conference.

3) Freshman, J.D. 1977. Public Works Committee United State, Senate, Washington D.C. 1971 A Perspective on Land as a Waste Management Alternative. Pages 1-8. Proceeding of the 1976, Cornell Agricultural Waste Management Alternative Conference.

4) Grier, H. E., W. Burton, and S. C. Tiwari. 1977. Overland Recycling System for Animal Waste Treatment. Land As A Waste Management Alternative. p. 693-702 Proceedings of the 1976 Cornell Agricultural Waste Management Conference.

5) Haider, K., F.F. Groblinghoft, T. Beck, H.R Sultan, R Hempfling and H.D. Ludermann. 1990. Influence of Soil Management Practices on the Organic Matter Structure and the Biochemical turnover of Plant Residues. p. 79-91. Advances in Soil Organic Matter Research: The Impact on Agriculture and the Environment.

6) Lamarca, Carlos Crovetto, 1996, Stubble over the soil. American Society of Agronomy Inc., 677 South Segoe Road, Madison, WI, 53711, USA.

7) Soil Quality Information Sheet. Soil Quality Introduction. USDA NRCS April 1996. Prepared by National Soil Survey Center in Cooperation with the Soil Quality Institute, NRCS, and USDA.

INDEX

About the Author

Suresh C. Tiwari was born in India in year 1932. He graduated with Intermediate in Science in year 1948. He joined Agricultural College, Sabour, Bihar, India in 1951and graduated with B, S, degree in Agriculture in year 1954.

He joined Bihar Agriculture Research Service in year 1954. Since 1954 to 1964 he worked in various capacities including Senior Soil Survey Assistant, Demonstrator in Soil Teaching Laboratory, as well as a Subject Matter Specialist in Soils.

In year 1964 he joined The University of Georgia in Athens for his Graduate Studies. He received his M. S. in year 1965 and his Ph.D. in 1969 from The University Of Georgia.

In year 1969 he joined Alcorn State University as an Associate Professor in The Department of Agriculture. Later he was promoted to full Professorship and continued up to year 1997 until he finally retired. During that period of academic service of 28 years he, on the basis of his research could produce sixty-two scientific publications.

Dr Tiwari holds the life memberships of American Society of Agronomy, Soil and Water conservation, Sigma Xi, Phi Kappa Phi, Gamma Sigma Delta and Mississippi Academy of Sciences

His present involvement is managing his Tree Farm and his small Garden in the rural Area of South West Mississippi.